To Dick, the inspiration, and Steven, the catalyst, and
to all the young ecological engineers – especially my
friend Chris M., who left us much too early.

For more information write:

Ecotone Publishing

721 NW Ninth Avenue, Suite 195

Portland, Oregon 97209

Author: David R. Macaulay

Book Design: softfirm

Illustrations: Glumac (unless otherwise noted)

Edited by: Fred McLennan

Library of Congress Control Number: 2010936984

Library of Congress Cataloging-in Publication Data

ISBN 978-0-9827749-0-8

1. Engineering 2. Architecture 3. Environment

First Edition

Printed in Canada on Reincarnation Matte paper – one hundred percent recycled content, Processed Chlorine-Free, using vegetable-based ink.

THE ECOLOGICAL ENGINEER – GLUMAC

TABLE OF CONTENTS

ABOUT ECOTONE PUBLISHING

THE GREEN BUILDING PUBLISHER

Ecotone Publishing was founded in 2004 in Kansas City by green building experts, Jason F. McLennan and Bob Berkebile. Pioneers and leaders of the burgeoning green building industry, Jason and Bob recognized a real need for educational resources that could assist industry practitioners and design students in achieving a transformation from traditional building practices toward sustainable design. Thus Ecotone Publishing became the first book publisher to focus solely on green architecture and design.

Ecotone Publishing is dedicated to meeting the growing demand for authoritative and accessible books on sustainable design, materials selection and building techniques in North America and beyond. Now located in the Cascadia region, Ecotone and Cascadia Green Building Council have joined forces within the International Living Future Institute, in pursuit of a united mission to transform the built environment to one that is socially just, culturally rich and ecologically restorative. Instilled with the same pioneering spirit from that which the company was born, Ecotone searches out inspiring firms, projects, people and vital trends that can lead the design industry to transformational change toward a healthier planet.

Our dedication to producing leading-edge resources for design professionals is in step with our desire to operate all aspects of our business with the environment in mind. Ecotone institutes a variety of green strategies, including the printing of its books on recycled content paper with vegetable-based inks, in facilities powered by electricity from sustainable sources, offset with certified renewable energy certificates. Ecotone utilizes recyclable packaging and selects carriers that place a high importance on conservation, efficiency and aggressive CO_2 reduction. Our commitment to the environment and to a sustainable future is the driving force behind every action we undertake.

ENVIRONMENTAL BENEFITS STATEMENT

Ecotone Publishing saved the following resources by printing the pages of this book on chlorine free paper made with 60% post-consumer waste.

TREES	WATER	ENERGY	SOLID WASTE	GREENHOUSE GASES
24	**10,858**	**10**	**688**	**2,408**
FULLY GROWN	GALLONS	MILLION BTUs	POUNDS	POUNDS

Environmental impact estimates were made using the Environmental Paper Network Paper Calculator. For more information visit www.papercalculator.org.

ECOTONE AND GLUMAC

Ecotone is extremely proud to serve as the chronicler of one of the finest engineering firms in North America. This compelling book provides a noteworthy sustainable design education - disseminating powerful ideas and celebrating innovations, practices and principles that deserve special recognition and study.

The Glumac team, as evidenced in the projects highlighted in this publication, is significantly helping to advance the green building movement. We are honored to work with this fine collective of leading sustainability thinkers and practitioners who are dedicated to moving the industry forward.

COLOPHON

Reflecting the environmental ethos of Ecotone, we selected environmentally-responsible materials for the production of this book. Reincarnation Matte paper is one hundred percent recycled, manufactured from sixty percent Post-Consumer Waste fiber (PCW), is Process Chlorine Free (PCF) and utilizes electricity that is offset with Green-e® certified renewable energy certificates. Reincarnation Matte paper is ancient forest friendly. Utilization reduces solid waste from entering landfill sites, and uses less water in manufacturing than conventional paper, decreasing air and water pollution. Ecotone also chose vegetable-based ink to compliment the paper selection. Compared to petroleum inks, vegetable-based inks release less than twenty percent of the mass of volatile organic chemicals and are more readily recycled.

FOREWORD

BY ED FRIEDRICHS

Design collaboration among architects, engineers, and contractors has simply not been the norm. Architects have generally looked upon engineers and engineering services with little interest, as things that need to be dealt with to "get my building built." To a significant extent, I believe that engineers have brought on this antipathy by avoiding creative participation in the design process. They have either been intimidated by what they considered to be arrogant architects or they were just not interested in sharing in the kind of creative joy architects find in their work. To these engineers, structure was only a means to hold up the building and mechanical, electrical, and plumbing systems were like young children in a Victorian household, better when not seen or heard. The same rancor also holds true for the relationships that architects and engineers have had with contractors, who have recounted many stories about design professionals who have little knowledge and even less concern with how things actually get built.

Throughout my career, I found that it was the exceptional engineer who would roll up his or her sleeves and brainstorm with me to advance my work and make it better. Most engineers were too uncomfortable and ineffective as communicators to contribute meaningfully to the design process. Yet, in those infrequent moments when a robust dialogue began to yield design results that surprised us, I felt that I was finally able to accomplish something worthy of being called architecture.

My experience with Glumac over the years has been the exception to the usual design process. In my early years as an architect, I had no delusions about my naïveté concerning engineering systems or construction methods. Living in San Francisco and being involved in the design of many buildings and interiors within walking distance of my office afforded me the best educational opportunity possible. I would hang out at the construction sites, watching and asking questions about how things got built and how we could better portray our intentions through our drawings. But the real treat came from working with the handful of engineers whose extensive practical knowledge and creativity helped me improve my work.

Dick Glumac, who had just started his firm, was one of those extraordinary individuals who always had the time and patience to share what he knew and who was never bashful about guiding me to an ingenious and innovative solution. As I have watched the firm grow and evolve, especially in the last seven years as a member of its board, I have been fascinated by the ability of the Glumac principals to sustain Dick's spirit in the collaborative work they deliver while working with a body of inventive and demanding clients.

We live in a world of ever-finer tolerances, increasingly restrictive codes and a laser focus on energy conservation that freezes many people into a mode of compliance, throwing creativity out the window. It is the rare professional who is willing to take the risk of innovation and to deviate from prescribed solutions. Over the years I have heard endless hype about exotic approaches to sustainability only to find, upon follow-up, that the designs are not performing as intended. However, I have been most impressed with Glumac's ability to develop solutions that are far advanced from textbook or code-governed approaches. Glumac engineers apply a perseverance that ensures what they design performs to their commitments. Their buildings really work! They consistently deliver creative engineering solutions that result in high performance and reduced energy consumption. Glumac works collaboratively with other professional teams to make certain that the mechanical, electrical, lighting and plumbing solutions enhance the structure and architectural envelope as well as the building performance. Great building design is not achievable today, in the era of restrictive regulatory environments, demanding budgets and elevated client expectations, without a collaborative effort involving architects, engineers and contractors, an approach that Glumac embraces to its core.

Glumac is further set apart from other engineering firms because of its unique perspective: a deep commitment to the effects their work has on human comfort and performance. Architects, whether they are willing to admit it or not, are dependent on their engineering consultants for the success of their work. These consultants provide the "hidden" stuff that makes a building feel wonderful or dreadful; that delivers air at just the right temperature and humidity, without draft, odor or dust and with a minimal expenditure of energy; that makes things on desks visible, without glare or eye strain; that helps a patient in a hospital heal more quickly and reduces the risk of infection; that makes the products in a store "jump off the shelf"; that allows a theater to transport people to another world. Poor performance in any engineering solution can undermine or destroy the work of even the most brilliant architect.

This book presents a limited selection of the entire body of Glumac's work and it showcases the design principles applied to everything they do. It expresses the reason I am honored to serve on their board. I am thrilled to see a book in print that celebrates Glumac's dedication to filling the world with great places for us all.

INTRODUCTION

BY STEVEN STRAUS

What makes a great building? Quality means different things to different people. Occupants want a spectacular space that is filled with natural light and fresh air, and is visually pleasing and comfortable. The owner wants a good building that is cost-effective, including both first and operating costs. To the maintenance staff, a great structure means it can be easily maintained with minimal occupant complaints. The architect might strive for acknowledgement from the community that the project is considered "Great Architecture," that helps win future commissions. For the contractor, a quality project is easily and safely constructed from local materials, with everyone on the team making a fair profit. At Glumac, we are focused on achieving all of these goals.

I first started in the green building business while attending UC Berkeley during the oil embargo of the late 1970s. It seemed many of the problems our country and the world were facing were related to dependence on foreign oil. I became passionate about solar energy, which evolved into passive solar designs for buildings. Many of the original hot water systems were plagued with problems such as failed and inefficient panels, and inadequate freeze protection. Great intentions mean nothing unless the system works.

This same idea applies to buildings. Many building projects have ambitious goals but fall short on delivering energy efficiency, comfort and rate of return on the investment. Creating great buildings is challenging with so many competing interests. Developments are often lead by a project manager who is rated solely on achieving the project within a set budget and schedule, regardless of maintenance and operating costs or sustainable features. Unlike past engineers, who worked in silos to create systems within the developed design concept, today we provide valuable information to the team about how shape, orientation and integrated systems interact to achieve a spectacular building. This collaborative work is enjoyable and results in better structures.

I have often heard members of the industry refer to building service engineering work as more of an art than a science. There is art in the design and architecture of the building, but to the ecological engineer, there is science as well. It is extraordinarily complex to predict the comfort or illumination of a space. Historically, it has not been cost-effective to conduct the detailed analyses of building structures and systems. However, new software technology has become available, including energy modeling, computational fluid dynamics, and daylight harvesting. These technologies allow us to provide a detailed examination of buildings that were not available ten years ago. More importantly, commissioning has also gained acceptance over the last decade, which provides feedback to designers about what is working and what is not to ensure that new complex systems are properly installed and operating.

For the last forty years, Glumac has been focused on helping design and commission "Green Buildings that Work."

ACKNOWLEDGEMENTS

When I first met Steven Straus, Glumac's president, he asked that *The Ecological Engineer* emphasize "green buildings that work" as one of its primary themes. "If we accomplish nothing else," he said, "it would be important to present – in black-and-white, hard data – exactly how and why these sustainable MEP designs save energy and water and create healthier, more comfortable places to live and work. And in telling that story, and Glumac's ongoing evolution as a firm, we strive to be factual, visual, and as concise as possible." In fact, he shared one of his favorite quotes: "I have made this letter longer than usual, only because I have not had time to make it shorter" – often attributed to Mark Twain, but actually (and more appropriately perhaps) written in a letter by the 17th-century French philosopher and mathematician, Blaise Pascal.

Since my first introduction to Steven and the firm in April 2009, I have had the pleasure of spending many, many hours with Glumac engineers, learning first hand about daylighting and building controls, rainwater catchment schemes, night flushing and underfloor air distribution, outside air economizers, chilled beams, switchable glass façades, photovoltaic arrays, urban wind turbines, and the untapped potential of fuel cells. I have also learned that in addition to design expertise and innovative problem-solving, some of the qualities most prized here are not necessarily those most people would expect when hiring young engineers: an appreciation of art and architecture (and music and literature) and the ability to collaborate and communicate exceedingly well. Capturing these traits firm-wide, along with its 40-year history and in-depth profiles of some of North America's most sustainably designed buildings – all became an exciting challenge in writing this book, and not one accomplished in a "short letter." Still, like Glumac's own signature designs, this narrative strives for a "less is more" approach, succinct but comprehensive – and is certainly enhanced by its collection of terrific photography, custom illustrations, and the as-always elegant book design and layout by Ecotone Publishing and its team.

Many individuals have contributed important perspectives on the firm's history and projects. First and foremost, it has been an honor to get acquainted with Dick Glumac, the firm's founder, a Serbian immigrant and once a world-class polo player who struck out on his own after working in California's aerospace industry of the early 1960s, and then consulting on San Francisco's groundbreaking Embarcadero Center. Even at eighty years old (and still running the local "Bay to Breakers" road race), Dick's intense eye for detail, his energy and his entrepreneurial spirit come through, as well as his insistence on maintaining high ethical standards and a strong business sense on every project, large or small. It is no wonder, then, this engineering practice has earned the trust and respect of some of the world's top architects, general contractors and developers.

This journey with Glumac has also introduced me to dozens of Glumac principals, associates and veteran designers as well as new hires from Seattle to Irvine and every office in between. In Portland, I toured award-winning buildings with Kirk Davis and James Thomas, and again with Rick Thomas in San Francisco and Scott Vollmoeller in Seattle. I spent valuable time with Edwin Lee, David Summers and Jacob Chan in Los Angeles, then Irvine's Richard Holzer and Herb Knieriem of the Sacramento office. In particular, Bob Schroeder and Leonard Klein gave their experienced insights on Glumac processes and methodologies, particularly the firm's success with integrated design and innovative tools to resolve and communicate engineering challenges. Steve Carroll shared his observations on commissioning, along with Carlos Inclán on daylighting studies, Jennifer Streb on lighting design, Mike Steinmann on green data centers, Hui Jin on CFD modeling, and Tommy Xing on Revit. I learned why some have been with the firm for decades and why others, like Lauren Kuntz and Christina Guichard, chose to further their careers here. Likewise, it has been a pleasure to learn from those young engineers like Rob Schnare, Brian Berg, Christopher Kerr, Grayson Hart, Jennifer Riehl, Dana Troy, Max Wilson – and everyone else who took part in the amazing leadership retreat held at the Portland Art Museum in June, including Angela Sheehan, Glumac's CFO.

Also, I believe it is important to recognize Glumac's clients, not simply as a gesture but because they influence and are influenced by the MEP design thinking so critical to the ever-growing list of projects that continue to transform the commercial, institutional and public built environment. Among those contributing valuable observations for *The Ecological Engineer - Glumac* were Gerding Edlen, ZGF Architects, Gensler, the GSA, Oregon State University, and the Portland Center Stage.

Trying to keep up with Steven – whether on foot, by car or by plane, around Portland and up and down the West Coast (visiting offices in four cities in three days) – is no easy task, and for that I thank Verna Triller for her persistence and scheduling wizardry to keep us moving forward. Special thanks also goes to Christine Hamilton, who played an essential role in completion of *The Ecological Engineer – Glumac*, by tracking down key Glumac staff, reviewing text, locating and negotiating photography, coordinating illustrations, and artfully tackling just about any task necessary to make this book happen. Thanks as well to Glumac's marketing team, including Whitney Peabody, Kimberly Tran, Winnie Rich and Michele Reesink, for assisting with interviews, photography and fact-checking. And to Willie Dean, a fresh new face at Glumac, for his excellent original illustrations of MEP designs that explain how many of their groundbreaking systems really work.

Thank you once again to the Ecotone Publishing team – Jason, Michael, Fred and Erin – for your encouragement, guidance and keen eye for detail in helping to make this second book in *The Ecological Engineer* series a reality. Finally, I want to thank my wife Lanie and kids Eli and Susannah for their patience and humor while giving me the space and time needed to research and write such an important chapter in the evolution of the green building industry.

Dave Macaulay, August 2011

PART I

THE PRACTICE

BAY AREA BEGINNINGS:
Work on San Francisco's
Embarcadero Center
provided the time (1967),
place and opportunity
for Dick Glumac to help
establish one of the city's
most iconic mixed-used
developments.

CHAPTER ONE:

YESTERDAY

OF MEMORY, CHANGE, RENEWAL AND PROMISE

"No one thought they'd get rich here or design the world's tallest building or even the biggest, but there's always been a continuity of wanting to solve problems, of serving other people and doing the right thing. I would like to be able to walk around or take my grandkids someplace and say 'I did that' and be proud of it – having something sustainable, something that lasts, something with meaning that survives over time."

— *Rick Thomas*
Principal, Glumac

"Overshadowing everything else is the question of conservation of natural resources. For how much longer are we going to waste...resources to save first-cost only of buildings?"

— S.R. Lewis, ASHVE President, 1914

1982

On a clear spring day in 1982, Dick Glumac surveyed the finishing touches of Embarcadero Center from his office window on the 15th floor. Spanning five city blocks and four million square feet next to San Francisco's Financial District, the shopping/office complex stood as one of the largest developments of its kind in the United States. He recalled a time, just a few years earlier, when his young firm was first hired to lead the mechanical and electrical design of the latest tower, Four Embarcadero Center: "I had done a lot of projects over fifteen years but never been out on my own. So I wondered, 'Do I have enough experience? Do I have enough knowledge? How do you design a 45-story building?' Soon afterwards, I never had any more doubt – it just came so spontaneously from then on."

2001

By September 2001, Steven Straus had served as Glumac's new president for just under a year. Representatives of the Jim Collins Institute (authors of the book *Good to Great*) had been invited to lead discussions on the future of the firm at its annual retreat in San Diego. There, they posed questions like "What are you passionate about? What can you be the best in the world at? Should every one of you really be sitting around the table?" A twenty-year veteran of the firm, Straus listened carefully as, one by one, each principal stressed values such as technical expertise and client service. "We actually started debating, arguing, screaming," he remembered, "and we built consensus that sustainability was important, something we could all agree upon. So that was the moment, right then. It finally dawned on me that we would make a commitment to moving forward on sustainability."

EMBARCADERO CENTER: With Dick Glumac and his young consulting engineering firm leading electrical and mechanical designs for towers One through Four, the complex continued its expansion towards the waterfront until complete in 1983.

A RISING NEW SKYLINE: With construction underway for One Embarcadero (bottom) and, later, Four Embarcadero (top), the commercial complex of office towers and hotels would transform San Francisco's financial district into a vibrant city within a city – and serve as home to a growing Glumac & Associates during the 1970s and 1980s.

"I was taught as a young engineer that my obligation is to the people who are going to live in the buildings we design, that these buildings have to perform. For thirty-three years our offices were situated in the very buildings we designed – no escaping it, anything they felt, we felt too."

— *Dick Glumac, Founder, Glumac & Associates*

2010

Moving quickly between the 3-D BIM views on his desktop monitor and over-sized schematics on a nearby table, David Summers orchestrated several key changes over the phone with the architect and client of his newest project. Upon its completion in 2013, the Los Angeles Trade Tech College Construction and Technology Building will implement many of the cutting-edge technologies that first attracted him to Glumac. The design, the process, the anticipated outcome for this state-of-the-art building all exemplify Glumac's L.A. office. One of the newest principals in the firm, his primary role on this multi-year project again comes down to "speaking the same language as the contractor to make it all happen. That's the difference. Ultimately, we're trying to match engineers with projects and clients who share a similar passion and interest in sustainability."

"THE THREE O'CLOCK WIND"

A walk just three blocks in any direction from Glumac's downtown San Francisco office – along California, Davis, Clay or Sacramento streets – offers a virtual timeline of MEP projects completed by the firm. For forty years, its designers have played an intimate role in the core-and-shell design and tenant improvements revitalizing some of the city's most iconic buildings within the Financial and Waterfront districts.

Yet the '"keystone" in Glumac's history is undoubtedly the Embarcadero Center (EC). Designed and constructed between 1967 and 1983, this groundbreaking development offered an early opportunity for the firm to make its mark with energy efficiency and innovative HVAC systems. First as an employee of Buonaccorsi & Associates and later as a consulting engineer, Dick Glumac managed the mechanical design of what was briefly San Francisco's tallest building. At 45 floors, One Embarcadero became the first of a series of towers and retail spaces stretching along Sacramento Street to the waterfront. Owner/architect John Portman of Atlanta wanted this international-style building to serve as a showpiece, featuring prime tenant space with floor-to-ceiling glass windows. Dick and his team recommended induction units (active chilled beams) for HVAC throughout. However, this presented a challenge for Portman: the units, one-foot-wide (0.3 meters) and positioned one foot from each window, would also obstruct views and reduce leasable space. For those reasons Portman requested that the units be placed above the ceiling. Glumac's design worked – the first overhead induction system west of the Mississippi – and still performs well into its fifth decade.

By 1971, Dick had left Buonaccorsi to launch his own consulting firm, designing systems for small offices and restaurants, schools, and an army hospital near The Presidio. When Portman's

EC Properties company (as developer) chose not to renew Buonaccorsi's contract on One Embarcadero, they asked Dick to complete the core-and-shell designs and tenant work for the building. He accepted. That year, he also hired his first engineer, Peter Mele, an acoustical specialist, to work on a new library for the Twelfth Naval District in Monterey. And he opened his first office, just 600 square-feet (56 square-meters), on the 21st floor of One Embarcadero as "Glumac & Associates." Over the next nine years, his small firm would occupy another six spaces throughout the tower. Here too, he would come to know the building well: appreciating its views west to the Farallon Islands and north to The Piers – even to understand the effects of "the 'three o'clock wind' pouring off the coast every afternoon as it hits this building."

The next year, 1972, Dick needed electrical design expertise. So he called upon Cal Webster, a former Buonaccorsi associate, and the firm became "Glumac & Webster." Two years later, Dart Rinefort, who had spent twenty years with Carrier Corporation, joined as the newest partner; he was followed by Rick Thomas from Skidmore Owings & Merrill's Chicago office. In 1977, the group formed a new corporation to reflect the new partnership, returning to the name "Glumac & Associates." As the engineering practice grew in size (to twenty employees) over the remainder of the decade, the projects and clients continued to grow as well through work with architects such as Gensler, EPR and KMD.

Dick had been involved only marginally with the designs for Two Embarcadero and Three Embarcadero, the next towers (each thirty-five stories) to rise in the expanding EC complex. Yet, in 1977 he got the call once again from Portman & Associates – this time to lead mechanical work on the final tower, Four Embarcadero, a mirror image of One but slightly larger and clearly the most energy-efficient building designed by Glumac up to that point. Instead of induction units, engineers specified a Variable Air Volume (VAV) re-heat system – featuring four 150 hp motors (110 kilowatts) and larger ducts to deliver half the cubic feet per minute (cfm) as the first building, with horsepower reduced by a factor of eight. Upon completion of this newest high-rise, Glumac's office took up residence on the 15th floor, remaining there for ten years.

TESTING THE WATERS

"You might say the Eighties were our high-rise years," said Rick Thomas. "At the time, they were popping up everywhere in San Francisco, and we did more of them probably than any other firm." First came the Pacific Lumber Building in 1982. With each new office tower, Glumac reinforced its reputation among a growing number of developers and architects as one of the city's most reliable, technically proficient building engineers. And the projects kept coming: 90 New Montgomery, 101 California, 71 Stevenson, and a major renovation for 901 Market at Fifth (on the National Register of Historic Places). Glumac designers also branched out into the emerging high-tech market, beginning work on data centers and several Stanford University laboratory buildings.

Given the Bay Area's temperate climate, Glumac consistently promoted the use of outside air to cool buildings. "We're blessed with mild weather on the West Coast," noted Thomas, "and most of the building load relates to cooling, not heating. Here, a building owner can literally turn off the chiller, run big fans and push air through the building. It's fresh and clean, saves energy and comes in off the ocean – and it's free cooling." Key to this concept was developing the "EconoMAX System," an air-side economizer that relies on blended air streams, and no chiller, to reduce HVAC energy costs while improving comfort and indoor air quality. (See "Bringing Outside Air In.")

Glumac & Associates welcomed 1983 by opening its first office outside of San Francisco. Steven Straus, still fresh from UC Berkeley with a degree in mechanical engineering, agreed to move north to Portland, Oregon. There, he would test the potential waters of the local construction market for Glumac. All partners put up money for the venture, and off he went: right into Portland's still depressed economy of the early 1980s. "Dick was shrewd and worked a bit of magic," said Straus. "He figured: what's the worst thing that could happen with a 25-year-old engineer eager to find new business? It was not a big investment and worth the risk." Gaining a foothold in the Portland market did not take long. Soon,

Glumac landed a job designing phone switch facilities for GTE, which "ended up being enough work for twenty people for a year," he added. Then came a big project for Sequent Computer Systems' manufacturing operations in Beaverton, Oregon, followed by design for the Mentor Graphics World Campus in Wilsonville, Oregon.

The next year, 1984, Glumac opened its third office – this time in Sacramento. Al Heitz, then working out of the San Francisco office, had suggested a presence for the company in California's capital. Dick agreed. Shortly thereafter, they won a proposal to remodel the antiquated heating and ventilating systems of the San Quentin Prison, a project that lasted for the next five years. Using what Dick Glumac referred to as a "practical 'Motel 6' approach" to mechanical and sprinkler design, the firm cut the budgeted $52 million cost of new systems down to approximately $24 million and created more humane conditions for the 4,800 prison cells in the facility.

Before the end of the decade, Glumac had cemented its reputation for high quality tenant improvements throughout San Francisco, completed Apple Computer's new campus in Cupertino, California, and increasingly ventured into high-tech and laboratory design. In addition, John Portman had selected the firm as the mechanical and electrical engineers of record for the new Embarcadero West – the final piece of the city's successful Embarcadero Center.

MENTOR GRAPHICS, 1983: Another milestone for Glumac's growing practice, innovative designs at the Mentor Graphics World Campus became the latest in a series of commissions for new electronic manufacturing facilities along the West Coast.

"At that point we decided it would be okay to do any kind of job, whether hospitals or high-rises or anything else. But they had to be sustainable projects – otherwise the client should probably hire somebody else, not Glumac. And that was, I think, the big turning point."

— *Steven Straus, President, Glumac*

JUST DESIGN IT

The 1990s brought more opportunities for Glumac to demonstrate its expertise with high-performance buildings further up and down the West Coast. In 1990, Richard Holzer joined the firm, establishing Glumac-Irvine to spearhead projects in Orange County and across Southern California. By 1996, Glumac-Seattle was added as well. While a focus on these new regions had been planned out carefully, the partners also agreed that these offices, like others outside of San Francisco, should be run independently. "Dick encouraged this entrepreneurial spirit by letting young people open up offices, engineers without much money to invest on their own but who were willing to put in sweat equity," observed Straus. "That's how all of these got started."

Research facilities, high-tech manufacturing, the dot com boom – for Glumac, the private and public sector work came fast and furious in designing data centers, server rooms, office spaces and laboratories from Seattle to San Diego. Projects included the Microbiology Lab and Gates Computer Science Building at Stanford University, the Earth & Marine Sciences Building at UC Santa Cruz, and an increasing number of clients in Silicon Valley like Digital Equipment Corporation as well as Hewlett-Packard and Digital Island (co-location facilities) near Portland.

Meanwhile, Acoustical Consultants Inc. (ACI), a stand-alone firm owned by Dick Glumac, also took on a series of large projects for the Lawrence Livermore National Laboratory in Berkeley, California. Dick had acquired ACI in the early 1980s to provide dust and noise control, vibration testing and other acoustic engineering solutions. In turn, this led to extensive work in California on clean rooms, computer chip manufacturing facilities and more for HP, Intel and AMD. On behalf of ACI, Dick and his team traveled all over the world, wherever an electronics facility was planned: to Japan, China, Singapore, Israel, and later to Scotland for Motorola and India for India Telephone.

Back in San Francisco, the firm continued to focus on high-rises with many of the city's largest commercial property owners. These projects included the renovations of 601 Montgomery and One Market Plaza and a complete remodel/ retrofit for 211 Main, the new headquarters for Charles Schwab's IT operations. Glumac also remained the engineer of choice for the majority of tenant improvement work across all four buildings of the Embarcadero Center and Embarcadero West. "It was my lucky break there," recalled Dick Glumac. "When you're filling up forty-one floors of a certain building, you meet about twenty different architects – and all those guys came back to me: always another space, another building for MEP design."

In this pre-LEED® era with the U.S. Green Building Council still emerging, Glumac had already gained a reputation in the market for its energy-conscious designs and ideas about air-side HVAC. In many ways, these innovations dated back to the mid-1970s and the focus on energy conservation during the Carter Administration – ideas then embraced by California's Governor Jerry Brown who rolled out a series of progressive energy-efficiency mandates. Steven Straus believed that Title 24 (introduced in 1978 and considered by many to be the strictest energy code in the U.S.) gave Glumac's building designs a head start on the rest of the nation: "Although energy consumption in the state has increased since then, it's up only slightly compared to actual population growth. In other words, California is one of the few states that's reduced its energy consumption per capita every year because of all the energy conservation measures put into place more than thirty years ago."

Likewise, Oregon's tax codes accelerated greener designs, offering some of the nation's best energy rebates following passage of the State Energy Efficiency Design (SEED) program in 1991. SEED required that all state facilities be designed, constructed, renovated and operated to minimize the use of nonrenewable energy resources and serve as models of energy efficiency. At Glumac, this requirement led immediately to a variety of projects for the General Services Administration and the State of Oregon.

Greater Portland's private sector in the mid-1990s also began to express a desire for healthier, more environmentally-friendly buildings and spaces to reflect its progressive – and growing – urban identity. As part of a project team that included architects Thompson Vaivoda & Associates and Kiewit Construction Company, Glumac embarked on a multi-year initiative to design new buildings for the Nike, Inc. campus in Beaverton, Oregon. In all, the firm delivered mechanical, electrical and energy engineering services for eight buildings, including the Ken Griffey Jr. Building, the Pete Sampras Building and the Nolan Ryan Building. Completed in 1999, each of these signature structures featured a high-performance thermal envelope, occupancy sensors for HVAC and lighting, and a radiant slab for the cafeterias. Closer to home in Portland's Pearl District, Glumac led mechanical design for the new corporate headquarters of Wieden+Kennedy, one of the largest independent advertising agencies in the world. This 1999 historic renovation optimized daylit interior spaces and natural heating, cooling and ventilation to create a comfortable working environment for employees and visitors.

Whether mechanical/electrical innovations refined over decades or newer sustainable concepts, the designs by Glumac at Nike and Stanford and within downtown Portland's Fox Tower increasingly set the stage for bigger things to come.

THINKING INSIDE THE BUILDING

The firm began 2000 full of promise and with a new name: simply, "Glumac." For more than ten years, the principals had been discussing the pros and cons of merging their five offices into a single corporation. Those against the idea focused on the benefits of keeping an independent, entrepreneurial, grassroots approach to projects without the bureaucracy of a centralized headquarters. Those in favor argued for increased efficiency and collaboration between offices. "But I think the number one advantage," recalled Straus, "was that it would provide better service to clients, a seamless experience across offices, whether in Portland or Sacramento or anywhere else."

On January 1, 2000, the merger was complete. Dick Glumac remained as President until February 2001, when he handed over the reins to Steven Straus while Rick Thomas became Glumac's CFO. As chairman of the board, Dick still works out of the San Francisco office every day, focusing his time and efforts primarily on acoustical design as Dick Glumac Consulting Engineers.

In representing the next generation of leadership, Straus began his search for a new theme, an organizing principle around which the company could leverage its strengths with an eye to the future. He wanted to re-brand Glumac. Following a series of meetings and guest speakers, he invited consultants from the Jim Collins Institute to participate in the firm's 2001 management retreat at the Del Coronado Hotel in San Diego. Rick Thomas remembered a pivotal discussion:

"We'd read *Good to Great* as homework, and each of the nine principals around the table responded to the question, 'What makes you great?' To a person, we'd all believed that if you do your best, solve problems and take care of your clients then the money will come. Up to that point, the type of project we pursued didn't seem to matter. But during that meeting the word 'sustainability' emerged as a common theme."

This commitment to sustainability led to hiring WOW Branding, a Vancouver, BC agency, to help Glumac express its values and vision as "Thinking. Inside the Building." Added Thomas:

"Our new tagline and re-branding efforts made good business sense about where the market is going. And it fit the whole *Good to Great* ideal of what do we want as a legacy and our careers and life. I believe if you find something you're passionate about, you'll be successful at it."

Portland seemed the perfect place to embrace and promote sustainable design, especially given the state's generous tax incentives. So the Portland office led mechanical, energy and lighting design for Viridian Place (completed in 2000), the first

LEED Certified building in the Northwest. In 2001, its engineers completed the city's first underfloor air system at the new Tiger Woods Conference Center on the Nike campus, along with the Mia Hamm Building (high performance thermal envelope, daylight harvesting) and the Lance Armstrong Building (heat reclaim, daylighting and natural ventilation). During this period, Glumac had undertaken one of its most ambitious projects ever: extensive mechanical, electrical and sustainable design services for the new $200 million Brewery Blocks. Totaling 1.7 million square feet (158 thousand square meters), this five-block redevelopment in the Pearl District produced a vibrant mix of urban retail, office space and residential housing. Completed in stages during 2002 and 2003, all buildings achieved LEED Silver or LEED Gold. The final piece, in 2006, came with MEP design for an historic armory renovated as the LEED Platinum Gerding Theater.

With LEED experience concentrated in Portland (James Thomas became the firm's first LEED AP in August 2001), Glumac began to export this knowledge and several staff members to other offices. In Southern California, Glumac provided mechanical design for a new 625,000 square-foot (58,00 square-meter), three-story office at Toyota's USA Headquarters South Campus in Torrance; opened in 2003, it became the largest building yet to receive a LEED Gold rating. The next year, engineers completed utilities modernization and tenant improvements for Swinerton Inc.'s LEED Gold headquarters office in San Francisco. Expansion into new regions meant opening new offices in downtown Los Angeles (2005), Las Vegas (2007) and Silicon Valley (2007).

Still, the momentum for sustainably designed buildings has been strongest in Oregon. At Oregon State University in Corvallis, Glumac collaborated on a daylit, naturally-ventilated atrium for the Kelley Engineering Center – in 2004. It became the first LEED Gold academic engineering building in the United States. In 2005, the stunning Wayne L. Morse Federal Courthouse opened, featuring underfloor air and radiant heating and cooling as the first U.S. building of its type to achieve LEED Gold certification. In 2006, the Providence Newberg Medical Center became the first hospital in the United States to reach LEED Gold. And in November 2007, The Casey opened in downtown Portland as the first LEED Platinum condominium in the United States, thanks in part to Glumac's MEP services for schematic design and RFP development. Another breakthrough, the Glumac-designed high-rise, Twelve|West, opened a few blocks away in July 2009. Twelve|West featured high-performance glazing, extensive daylighting and radiant heating and cooling throughout. Designed to achieve two LEED Platinum ratings (NC and CI), it also earned honors as a 2010 AIA Top Ten Green award winner.

FOX TOWER, 2000: Rising high above Portland, the 27-story office building highlights use of chilled air, heat recovery and highly-efficient glazing developed through Glumac energy models.

MIA HAMM BUILDING, 2001: One of eight
buildings designed for Nike's headquarters
outside Portland, the company's
R&D facility celebrates a full range of
sustainable features: radiant heating and
cooling, daylighting, thermal mass and a
high-performance envelope.

NIKE WORLD CAMPUS, 1999-2004: At Nike's state-of-the-art corporate headquarters, Glumac delivered mechanical, electrical and energy engineering services for 1.1 million square feet of space space (102,000 square meters), introducing underfloor air systems and other breakthrough sustainable design elements. In addition to the Mia Hamm Building (interior pictured below and exterior pictured on the right), Glumac designers created strategies for the Ken Griffey Jr. Building (LEED Gold), the Nolan Ryan Building, the Pete Sampras Building, the Lance Armstrong Building, the Tiger Woods Conference Center, and the Jerry Rice Building.

VIRIDIAN PLACE, 2000: Creative lighting design, energy efficiency and high-performance HVAC systems all contribute to Portland's Viridian Place becoming the first LEED-certified building in the Northwest.

Glumac took this burgeoning LEED expertise to its own workspaces beginning in 2006. After thirty-three years at Embarcadero Center, the firm's San Francisco office moved into 150 California St., earning LEED CI Silver for tenant improvements and Energy Star certification for the building. That same year, the Portland office also achieved LEED Silver for its second and third floor space in the historic Dekum Building. In August 2007, the Sacramento staff took occupancy of a new LEED Platinum office, combining mechanical, electrical and plumbing design into a highly sustainable work environment. Glumac's Irvine office joined in, with an artful LEED CI Platinum renovation completed in March 2010 and featuring an energy-efficient design that emphasizes daylight and passive heating and cooling.

What comes next for the 40-year-old engineering firm? Advanced new designs for projects underway in Los Angeles, Salt Lake City and Shanghai. A new generation of engineers pressing to transform the very face of the built environment. One clear theme at Glumac is that its leaders often look to the values of the past to describe their hopes and dreams for the future.

For Dick Glumac, the firm's success has been the result of experience and continuously adopting new technology, tapping experts, completing work ahead of time – and always, the willingness to design and then observe a project's outcome in person. But mostly, he said, "it's been about honesty and integrity. I made my share of mistakes and admitted them right away and promised that I would fix them. Clients prefer to work with an engineer they know, someone they can trust."

An unwavering optimist, Steven Straus thinks back to an earlier time when as a young engineer he traveled every day over the Golden Gate Bridge between Sausalito and Glumac's office in San Francisco. He became fascinated with the history and politics of the bridge, its radical design and construction overseen by Joseph Strauss (no relation):

"No one had ever designed a bridge like that. He brought in an architect and didn't leave it strictly to the engineering, even built it using technology never imagined before. To me, it's the most amazing bridge in the world: there is something special about the way the sun comes up over it, the way the colors look. To understand that some engineer, against all odds, could get this bridge built, so adamant he was going to get the job done – that was really inspiring. In the same way, I've always wanted to help make Glumac a great company, not just a good company. No one has ever naturally ventilated a 70-story building, for example. But if you use great engineering principles and if you're tenacious, you can do it."

FIRST IN THE NATION: Glumac MEP designs played a significant role in several LEED firsts, including Providence Newberg Medical Center (2006), bottom left, the first hospital in the U.S. to be certified LEED Gold, and Portland's The Casey (2007), bottom right, the first condominium project in the U.S. to receive LEED Platinum.

DICK GLUMAC:

"UZDAJ SE U SE I U SVOJE KLJUSE"
("RELY ON YOURSELF AND YOUR MULE")

Mile "Dick" Glumac remembers swimming in the Danube like it was yesterday, the water cold, the sun overhead warming his shoulders as he headed back to shore at the edge of Belgrade. In that summer of 1946, just two years after his native Yugoslavia had been liberated from the Nazis, he was late for a mid-afternoon meeting of the Young Communist League (Savez komunisticke omladine Jugoslavije) downtown. It was a meeting he was dreading, a waste of time and filled with Party politics, and no place for a teenager on this hot, humid day.

Soon afterwards, he quit the group knowing, even then at the age of sixteen, his future was elsewhere, in America perhaps. He had begun studying English, occasionally visiting the local U.S. Information Center after school to read magazines like Life and Time. By the age of eighteen, he was working three jobs to support his family after his father had been arrested and sentenced to twenty years in prison for alleged collaboration with the enemy during World War II. By day, he worked as a mechanical designer and draftsman in a large Belgrade factory producing agricultural machinery and tool and die equipment for the military. At night, he practiced for hours with his water polo team, one of the country's youngest. Together, these paths would help him forge a way out of the Eastern Bloc – a way to the West.

THE WAR, A COWBOY, AND WATER POLO

Dick was born February 2, 1930, the oldest of four boys, to Milan and Lubica Glumac (pronounced 'Glue-matz', meaning "actor" in Serbian). Growing up in Zagreb, the capital of the Province of Croatia, he and his family had to flee when German forces invaded the country in 1941. Later reunited with his father, mother and brothers in Belgrade, they lived together in two rooms for the duration of the war until October 1944. Although his Christian name was Mile (Mi-lé), he had earned a new name after wearing hand-me-down clothing sent from his aunts in Gary, Indiana. He and his friends were regulars at the

Dick Glumac

movies, especially westerns starring Dick Foran. "I was the only kid in Belgrade who wore jeans in 1945," he recalled. So, at the age of fifteen he became "Dick" Glumac.

With Tito's Communist Party firmly in power, in 1945, he and others volunteered to pick corn on the plains of Yugoslavia, studying Marx and Engels at night and listening to speeches about proletariats. He began playing water polo and found he was good at it, a fast swimmer in the pool. From 1948, when his father was sent away, the family began to disintegrate: his mother died in 1950 and by 1951, his younger brother (by eleven months) was already serving in the artillery with his third brother recently drafted into the army. Now playing for the top state water polo team in Serbia, Dick and his teammates were sent to Innsbruck, Austria for an international meet. On June 11, 1951, he and seven others (including members of the swim team) defected, later taking a train to Salzburg: "I had an inkling that I would defect, but the Iron Curtain was really an 'iron curtain.' We had no clue what was out there."

DEFECTING TO CANADA

Safely in Salzburg, the group had a choice to make. To a man, they did not want to remain in Europe; however, there would be at least a three-year wait before they could immigrate to the United States. Other possible destinations included Australia, Argentina or Canada – "so we chose Canada, which we knew was close to US of A". With the help of the International Refugee Organization, they arrived in Ottawa a few months later. Yet, in a country still reeling from the post-war recession, the only jobs available were cutting lumber that winter in the forests of Canada, thirty miles by sleigh ride from Chalk River, Ontario. By the next fall, Dick enrolled at the University of Toronto to study mechanical engineering and with two of his fellow émigrés joined the school's varsity water polo team.

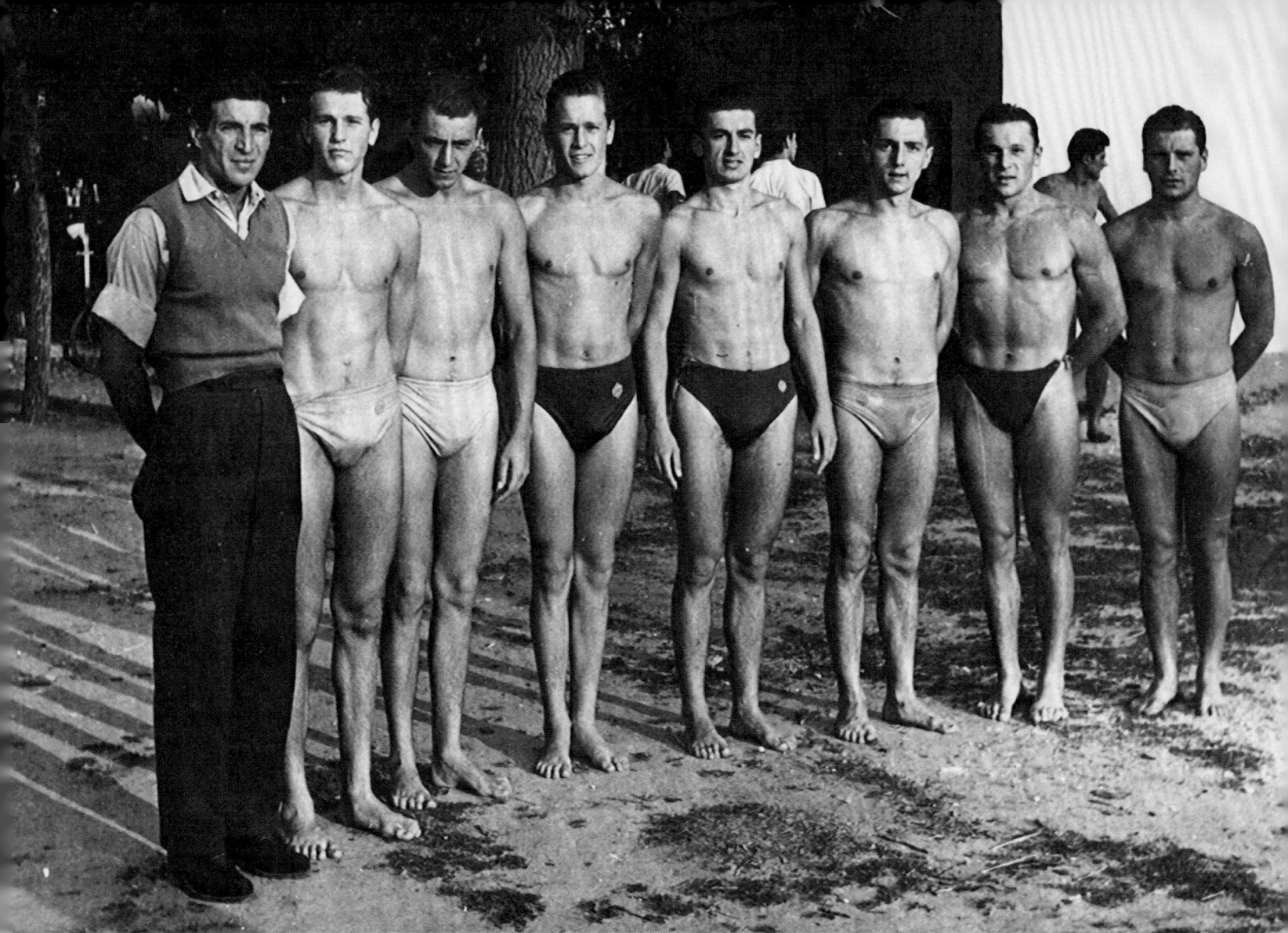

YUGLOSLAVIAN WATER POLO TEAM, 1951: Dick Glumac (second from right) and many of his teammates escaped the Iron Curtain during an international competition in Austria – and from there to Canada with dreams of living and working in the West.

Upon his graduation in 1956, he landed a position with Rybka, Smith and Ginsler in Toronto to focus on heating and ventilation, a return to the building design work he had first experienced while learning a trade in Yugoslavia. In search of still broader opportunities, he joined the construction firm Canadian Comstock as a project manager, transferring to Vancouver in 1959 to oversee a large cost-plus rush job for a new Kodak processing plant. The following year, because of a lengthy delay until his next assignment, Dick chose to take a few months off to visit Vera Cruz, Mexico, six thousand miles away (9,660 kilometers) – a slow trip from Vancouver to San Francisco, Reno, Los Angeles and finally the small town of Xalapa, Mexico. There, he stayed four months, studying the history and anthropology of Mexico and perfecting his Spanish. On his way back to Canada, he stopped in Los Angeles. It was September 1961. First, he purchased a brand new 1961 T-Bird 4-seater "with my last nickel." Then he opened up the Los Angeles Times, discovering more than forty pages of want ads for engineers at Bechtel, Martin Marietta and hundreds of other companies on the verge of unprecedented growth in the region's aerospace industry. "You called up anybody – 'I'm Dick Glumac, I'm looking for a mechanical engineering job' – and they'd hire you on the spot."

"HELP WANTED": DESIGNING FOR SPACE

Faced with many options as a young engineer, Dick was drawn to one of the suppliers working under a Rocketdyne (Division of North American Aviation) contract to build a test stand for engines for NASA's Saturn and Apollo rockets. As a designer for Enviratron Company, a test chamber manufacturer in Burbank, California, he participated in a project to develop the "Diffuser-Ejector" for testing the Rocketdyne's F-1 engine, still considered the most powerful single-chamber liquid-fueled rocket engine ever created. In addition, he started designing his first clean room, at Moffet Field in the early 1960s.

While Dick enjoyed his time in aerospace, he also wanted to return to building design, accepting a position with consulting mechanical engineers Hellman & Lober, Inc. There, he worked on several high-rises in greater Los Angeles, as well as high-end hotel/casinos in Reno and Las Vegas, including a major addition for the famous Sands Hotel. During this period, the 16-employee firm produced a large volume of designs, particularly for Arthur Froehlich & Associates, on Safeway supermarkets, the Imperial 400 and Astro Motel chains throughout the United States.

By the end of 1965, Dick and his family chose to leave smog-ridden L.A. to provide relief for his young daughter who was struggling with asthma. He considered three new locations in search of clean air: Honolulu, Fairbanks and San Francisco. Spotting an ad in an ASHRAE Journal, in December he flew to the Bay Area for an interview with Buonaccorsi & Associates, a 70-person firm specializing in hospitals, office buildings, schools and laboratories. He was hired immediately. On behalf of Buonaccorsi, over the next six years, he would design systems at Stanford University, the Hawaii Medical Center and elsewhere for prominent national architects that included Edward Durell Stone and Stone Marraccini Patterson.

BECOMING "GLUMAC & ASSOCIATES"

After taking a lead role for Buonaccorsi on expansion of the San Francisco International Airport, in 1967 Dick requested reassignment to head up mechanical design for a very different project downtown: a high-profile office/retail complex known as the Embarcadero Center (EC). Only recently had Buonaccorsi & Associates completed its work on the Walter Reed Army Medical Center in Washington, D.C. – a state-of-the-art facility and the firm's biggest project yet. By 1969, Dick had been promoted to a company director and continued working with some of the most talented mechanical and electrical engineers in the state, including Ed Kirchner. By day, he managed projects at One Embarcadero, the EC's first tower; and by night and on weekends, he pursued the smaller design jobs Buonaccorsi did not want, especially tenant improvements, small offices and schools. Soon, his "moonlighting" led to bigger contracts, with the Twelfth Naval District near San Bruno as well as the U.S. Army's Letterman Hospital. So, with new consulting gigs adding up and some money in the bank, he decided it was time to strike out on his own. "I had some apprehension, but not much – and there was very little planning," he said. "But I never questioned whether or not I made the right move." On October 1, 1971, Glumac & Associates was born.

Those decisions by Dick Glumac, to lead mechanical design at the Embarcadero Center and later to launch his own consulting engineering firm, set in motion a 40-year consulting relationship on one of the world's most famous urban developments that continues today.

THE DIFFUSER-EJECTOR: After earning his mechanical engineering degree, young Dick Glumac moved south to Los Angeles and began employment in the aerospace industry. There, he was hired as part of the project team to build test stands for NASA's Saturn and Apollo rocket engines.

THE CORVALLIS DECISION:
Glumac's ninth office, across the
street from the OSU campus in
Corvallis, provides a stronger
sustainable design presence for
the firm to clients in the Corvallis/
Eugene/Salem area – while
enabling its engineers to better
connect with young engineering
students passionate about green
buildings.

CHAPTER TWO:

TODAY

REIMAGINING THE ROAD TO THERE

"We have a responsibility, an important responsibility, to make sure we design green buildings that work. What we do really is transparent: you don't see the ductwork or the VAV boxes or even the light fixtures, but you feel it – you know that a room is comfortable, that the light is beautiful."

— Steven Straus
President, Glumac

"Get involved in something that you care so much about that you want to make it the greatest it can possibly be, not because of what you will get, but just because it can be done."

— Jim Collins, *Good to Great: Why Some Companies Make the Leap...and Others Don't*

Merging five offices into one corporation. A year-long self-examination of core values through Jim Collins' *Good to Great* process. A new focus on sustainability. The Glumac of today began to take shape in 2000, 2001 and 2002. "Ultimately, we concluded that we could be a better company as one corporation," recalled Steven Straus, who had just been named the firm's new president. "More importantly, we felt we could serve our clients much better together than as five separate organizations."

The outlines of this new corporate design, too, had emerged even earlier, with several unique opportunities to create high-performance building systems in and around Portland, including Glumac's first underfloor air/raised access floor for the City of Portland at 1900 SW Fourth Avenue. Work for Nike, the first green corporate campus of its time, had recently concluded, followed by breakthrough designs using underfloor air and daylighting for Wieden+Kennedy (see page 200). Also driving the push for greener buildings, the State of Oregon offered the best tax credits for energy efficiency in the nation. And combined with Portland's famous environmental ethic and culture of sustainability, Glumac started to design LEED® buildings: first one, then two, five, ten and twenty projects – expertise the firm could then export to its other offices along the West Coast.

So what makes Glumac – Glumac? Every successful company may be defined by its history and its body of work over time. And yet what also distinguishes this 40-year-old engineering firm is the human element, several generations of men and women with a shared passion for sustainable design. For Bob Schroeder, a 10-year veteran of Glumac who moved to the San Francisco office as a principal in September 2007, "First and foremost it's the people. It's a great bunch of people that you want to work for – and with. They're very collaborative. They care a lot about the projects from top

"We tend to attract people who are interested in sustainability already...clients come to us because they want their project to be sustainable, and designers join us to work on those projects. So I think we're just trying to create a place where we can put the engineers with the clients and the projects that all have a similar passion and interest."

— *David Summers, Principal, Glumac*

to bottom." Leonard Klein, a principal in Portland who has been with the firm since 1996, agreed, adding that Glumac designers consistently strive for innovation:

"The architect may not even be pushing it, but it's our job to be proactive and provide a compelling reason to integrate sustainable design into every building. Also, we have to stay current with the technology: what we designed even five years ago might be outdated. So we ask ourselves, 'What can I do to that envelope to reduce the window-to-wall ratio, to install more efficient glass, to open things up...maybe I don't even need mechanical cooling.' Those are the concepts we need to push."

What appeals most to Jennifer Streb, a Glumac employee, is Glumac's "cultural ability to give people room to grow and feel like nothing's really holding them back". She joined the Portland office in 2005 as an electrical engineer specializing in lighting design. "Personally, I do best in an environment where I feel like people believe in me, where they're behind me and enable me to challenge myself." A native of Seattle, David Summers joined Glumac-Portland in 2002, then helped to open a new Los Angeles office where he remains today as one of the firm's newest principals. He takes pride in promoting Glumac's breadth of sustainability knowledge, from concept through implementation:

"They build buildings differently in Europe: engineers are more focused on the schematics and concepts and then contractors figure out how to make it all work. What I noticed right away after joining Glumac was they understood the concepts as well as the nuts and bolts of how details need to work – controls, sequencing, how existing buildings operate – so we can talk the same language as the contractor. That's a big differentiator for us."

BEYOND *GOOD TO GREAT*

Commitment, confidence, relationships – these are just a few of the personal and corporate qualities that underlie Glumac's successes over the past decade. Talk with enough engineers across each of their offices, and these are the words and phrases they use over and over – each in its own way just as important to employees as the firm's reputation for innovative thinking and technical expertise. During that same period, Glumac leadership also agreed on twelve core values, collectively promoted to employees, clients, partners and prospective employees. It's not unusual to hear Steven Straus address these when describing their work, often distilling them down further into a few fundamentals that guide their practice and beliefs. "For starters," he told a group of young engineers recently, "we should always under-promise and over-deliver in terms of completing the work properly and ahead of time. And pay attention to detail," he added. "Design as if it were for your own building. Those are the elements that distinguish, not good companies, but great companies."

FACES OF GLUMAC AND
FREQUENT COLLABORATORS

DESIGN AT SEA: Glumac principals
continue to emphasize teamwork, fun and
the professional ethics embodied by Dick
Glumac and the firm's founders more than
forty years ago.

Straus also holds up founder Dick Glumac as a model of engineering ethics for his honest, pragmatic approach to business that has endured over the last forty years. Rick Thomas, San Francisco's Managing Principal who joined the firm in 1977, noted it is not unheard of for a designer to address a technical or financial challenge on a project by asking, "What would Dick do?" That speaks to Dick's philosophy of doing your best job, doing the right thing and following through with what you said you're going to do. Because we made a commitment to the client, we're not going to walk away from it. Future work will always come." Likewise, correcting a mistake, openly and honestly, is a theme echoed throughout the organization. "Standing behind the work that we do...that's a huge element of our beliefs and values," said Leonard Klein. "Projects don't always go perfectly. So the big question is, how do you respond? It requires working with the architect, the general contractor and the subcontractors to resolve the problem in a positive way that doesn't sacrifice any of the integrity of the systems you're presenting over to the owner."

Best practices at Glumac also include finding effective ways of communicating sustainable design concepts, issues and opportunities to clients, whether architects, developers, building owners or other decision-makers. As Bob Schroeder explained, "Clients are a diverse group, and they all speak different languages. One group may want just technical detail (the facilities staff) about how a chiller operates, while a developer may more interested in that chiller's efficiency per code. I think we do a good job of understanding our audience." Still, he expressed, presenting is a key part of the process for every stage of a project:

"People like to go through the journey, to go through the steps of getting from a blank sheet of paper to a set of viable options. Usually when that's successful, the team is at a level where they can understand the systems and the pros and cons of each; collectively then, we can come to a conclusion that this is the right system and move forward."

ONE CORPORATION, ONE GLUMAC

Another key difference for Glumac today is how staff members mobilize around a new project. Perhaps more than any other part of its evolution, the merger of offices into a single corporation, and later, a single profit center, has united the firm under one banner. And whether a job site is in the Sierra Foothills, downtown Seattle or along the San Diego coastline, the firm's basic approach remains the same. Generally, the Glumac office closest to the architect is responsible for project management, to facilitate better communication and to serve as a point-to-point contact for the entire project team. From there, each Glumac project manager assembles their internal mechanical, electrical, plumbing and commissioning team as

"Sometimes, the lower upfront capital cost is the way an owner or developer wants to go...so they pursue the cheaper solution. But we're continually trying to have some measure of sustainability in everything we do: as engineers, we try to be as cognizant of efficiencies and energy as possible."

— *Herb Knieriem, Principal, Glumac*

needed. And yet the overriding decision, said Herb Knieriem, Sacramento's managing principal, "is really based on what's best for the client."

Knieriem, who has been with Glumac since 1988, explained that "doing the right thing for the client and finding the right solution" has long been a matter of managing a project as efficiently as possible within the schedule and budget. That may mean pulling in team members from other offices who specialize in CFD modeling or daylighting – or tapping those with in-depth experience on a specific project type, such as auditoriums or retail stores. He elaborated further:

"One of our strengths is that we have expertise somewhere in the organization on pretty much anything you need. A young engineer in Sacramento may call somebody in Portland or San Francisco who has done the exact same project, had the same problem and has a solution. That's something we do pretty well: understanding how testing or codes apply to a particular building. It's amazing to me that usually within a half hour, they can come up with several dozen solutions."

Collaboration also occurs in unexpected ways when designers with vastly different backgrounds pool their experience, added Steven Straus, to arrive at out-of-the-box solutions – for example, integrating microelectronics design ideas within a laboratory

setting. "Courtrooms don't appear to have anything in common with prisons, so mechanical and electrical engineers sometime fail to realize how prisoner holding areas in courts could be designed and detailed. Also, OSU's Kelley Engineering Center, a higher education building, includes a data center as part of the program, so Glumac brought in our data center engineers to assist with the project."

Glumac's expertise across disciplines continues to grow as well, as engineers anticipate, refine and combine skill-sets to meet new market opportunities. Across all of its offices, the firm has identified "market sector champions" in eight areas, including healthcare, retail, high-rise office buildings, data centers, higher education, hospitality, government/institutional and commissioning. To its core MEP competencies, Glumac has also developed and/or added expertise in energy modeling, lighting design, rainwater systems design, and computational fluid dynamics (CFD) modeling.

GREEN BUILDINGS THAT WORK

Building commissioning holds great promise for Glumac. Every office currently includes staff – commissioning agents – dedicated solely to this process which Steven Straus sees as critical to creating "green buildings that work": "Determining what does or doesn't work, providing that feedback to the design group, then following up month after month and year after year

THE GREATER GOOD: Glumac's Habitat for
Humanity construction team in Portland,
Oregon, 2010.

ALL TOGETHER NOW: Glumac Los
Angeles Summer Family Picnic,
2007 – a celebration of life, growth,
accomplishments, friends and family.

to ensure every building is performing as intended – that's necessary to being a great design firm." Retro-commissioning, too, opens up an entirely new market with existing buildings. It begins with a detailed investigation of the high operating costs for a facility, then recommending green strategies to O&M staff: retrofitting plumbing fixtures, use of lighting and daylighting controls, tuning up HVAC systems, upgrading zone controls with a DDC interface, etcetera. "This process can take months," he said, "but you leave the owners and facilities people with a better product at the end."

Just as commissioning and CFD modeling represent two essential tools for Glumac today, its designers are continually discovering effective new ways to improve communication. Models, colorized drawings and diagrams all play a role in helping clients understand complex and often completely new design strategies for heating, cooling, air and water. Leonard Klein appreciates this freedom "to be innovative in how we present information." His efforts to devise a new energy metrics tool has helped speed up the decision-making process on several projects by exploring "what are viable options, what's the cost of those options, what's it going to take to maintain those options – and then have that dynamic discussion happen between the architect, the engineer, the general contractor, even the facilities people who will be operating the building after design and construction."

Still, the most powerful "tool" employed by Klein and other Glumac engineers is undoubtedly the integrated design process and team – a practice widely adopted throughout the firm which has a distinct impact on many of their most successful projects. Bob Schroeder knows from experience there is simply no substitute:

"It takes people around the table communicating back and forth. It's not like the old days, where you got the building plans, did your thing and returned at some other milestone to present your ideas. Decisions are made collaboratively. You need a team around the team who can think and collaborate and integrate... because when we're successful, we're integrating our systems into the architecture. It's all about creating great architecture. You create a great building and a great envelope and do all those things first; then, the system integration is easier."

Leadership at Glumac, too, takes many forms, valuing initiative, interaction and innovation as much as a diversity of technical skills in creating green buildings that work. It begins with hiring, with finding engineers who can collaborate both internally and externally – an ongoing theme for Straus and the rest of the managing principals: "Architects need engineering firms that appreciate art, music and architecture and who can respond to the challenges that will come up, creatively and flexibly. We've got to continue to recruit those people, bring them on and train them."

GLUMAC AT WORK: Whether collaborating at an office, listening in during a monthly Glumac University training session or volunteering at a Habitat for Humanity project, Glumac employees share a passion for sustainability – for pushing new concepts and out-of-the-box solutions to energy and comfort challenges.

Succession is another clear theme for the firm, preparing for change and the uncertainties – and opportunities – of the road ahead for a greener built environment. David Summers is among those representing Glumac's third generation who have been encouraged to take on a leadership role:

"The model at Glumac incentivizes good people, to attract and keep engineers who have that entrepreneurial spirit. It's a unique company in the sense that it's not huge, owned by thousands of employees; it's owned by the people who are the leaders now. At the end of the day, we want to be able to do the same thing we benefited from when we came up, to groom the next generation of potential leaders and give them all the opportunities to do everything they can do. We're always looking for the next leaders of the company."

ENGINEER AS MENTOR: Formal or informal, mentorship throughout Glumac pairs youth and experience to arrive at new ideas, new solutions based on sound engineering principles.

GREEN DATA CENTERS

CRITICAL FACILITIES: Creating high-performance, environmentally-friendly data centers – including designs for a Sun Microsystems facility (now owned by Oracle) in Santa Clara, California, completed in 2007. Glumac has contributed its design, construction and commissioning expertise to more than 300 data centers, collocation facilities, network operation centers and electronic switching stations.

What makes a data center *green*? As with many other project types, sustainable-design thinking here calls for high-energy efficiency as well as flexible approaches to cooling, water, power consumption and environmental controls. "It's an industry built around reliability," pointed out Mike Steinmann, a Principal in the Silicon Valley office, who serves as Glumac's market sector champion on data centers. "People lose their jobs if certain data centers go down, which causes many people to do the same thing they've always done because they know it works well. So that's driven construction projects – not operating costs or green design ideas."

To date, Glumac's "Critical Facilities Design Team" has worked on more than three hundred data centers, co-location facilities, network operation centers and electronic switching stations. Yet it is their pioneering concepts for Dell, Earthlink, Intel, Sun Microsystems and others that continue to garner attention from the data center industry nationwide. In collaboration with Dell, Glumac engineers developed an outside air system that, in combination with high-efficiency fanwall technology, could economize cooling 100 percent of the time in Austin's hot, humid climate. Through CFD modeling, they also devised a modular system capable of cooling up to 800 watts per square-foot (8.5 kilowatts per square-meter) – as compared to the standard of 150 to 300 watts per square-foot (1.6 to 3.2 kilowatts per square-meter)

for most facilities. For EarthLink's data center expansion in Pasadena, California, Glumac engineers delivered an innovative underfloor air distribution scheme to support a hot aisle/cold aisle arrangement. Based on the groundbreaking research of the Green Grid global consortium and the Rocky Mountain Institute, Glumac was able to demonstrate how server equipment may operate at higher temperatures (75° to 80°F or 23.9° to 26.7°C supply air) efficiently, allowing engineers to push the performance of air-side economizers even further. Many of these concepts were also deployed with stunning results for a high-density data center for Sun in Santa Clara, California (See "Sun Data Center," page 316).

"We've introduced a lot of new concepts for more efficient data centers, including the use of more outside air, and it's all starting to get more traction within the industry," said Steinmann. "We think you can increase a facility's reliability, lower your first cost and operating costs, and lower time to market with an air-cooled system. Therefore, it's now possible for more efficient equipment, with more megaflops per watt, to operate at a higher temperature and both higher and lower humidities than ever realized before."

A PASSION FOR SUSTAINABILITY

Rob Schnare and Christina Guichard share space within Glumac's downtown Portland office. Every day, all day, they collaborate on technical concepts and ideas about sustainability. As young mechanical engineers, they also have access to several of the firm's principals and senior designers just down the hall. And like so many "new hires" joining Glumac offices up and down the West Coast, their stories are familiar: *I always wanted to work for a firm that designs green buildings.*

Schnare joined the firm in 2006 – but not before touring much of North and South America. Immediately after graduating from Cal Poly, he "went straight to the mountains" – for snowboarding, rock climbing, skiing and backpacking – until his money ran out, and he joined an MEP firm. Assignments often took him to resorts in Colorado, California, Idaho and across the West. Still in search of adventure, he left his job for Argentina, where he spent four weeks climbing Aconcagua, the highest peak in the southern hemisphere, and then surfing up and down the coast of Chile. Schnare returned to the U.S., relocated to Portland and approached Glumac for work: "I'd been interested in LEED and sustainability early on, so I knew there was going to be a lot for me to learn."

Among his biggest accomplishments to date, Schnare completed work in 2009 on the LEED Platinum headquarters for Mercy Corps in Portland (see pages 232 – 235). Rob is especially proud that in addition to its daylit, naturally-ventilated interior, the design established new metrics for a building that could perform well beyond LEED. Guichard, likewise, has participated in several groundbreaking projects since joining the firm in 2009. These projects include an indoor water park and a new LEED Gold campus for Harvey Mudd College (Claremont, California), featuring radiant panels and designed to achieve future net-zero energy.

THIRD GENERATION: Glumac employees Christina Guichard and Rob Schnare collaborate on schematic designs for cutting-edge mechanical systems within a new academic center at California's Harvey Mudd College.

Guichard graduated with a mechanical engineering degree at North Carolina State, then worked for several companies but "kind of lacked inspiration." That led to volunteering for California's Leukemia & Lymphoma Society and, later, the opportunity to head up corporate development for the non-profit – a field she remains passionate about. Yet when she and her husband decided to move to Portland, it seemed the right time to transition back into engineering:

"It was hard to leave something where it's easy to see how you're making a difference. There is nothing more powerful, and that's changed my life forever. But it also gave me the courage to realize you only live this life once. To find a company with goals and values that lined up with mine, and for clients who actually cared about their carbon footprints...that was really important to me."

Her targeted job search led directly to Glumac. While Rob Schnare's path to Glumac was anything but conventional, he knows this is the ideal place to interact with the firm's leaders, his mentors, and gain a wide array of MEP experience. "They like having the younger generation around, because we have so much fire and passion about what we do. We're all learning things, so we're there to push them along, too."

NEW FRONTIERS: A progressive, comprehensive MEP approach to the Teaching and Learning Building at Harvey Mudd College (opening in 2013) symbolizes a new era of sustainable design thinking – a broad array of green elements on display and as a living laboratory on campus.

CHAPTER THREE:
TOMORROW

THE PROMISE OF PLACES YET TO BE REALIZED

"Fate loves the fearless."

—*James Russell Lowell*

"I always wanted to be an engineer, always knew I needed to do something I'd be passionate about. Sometimes you have to reach – beyond creating a green building – to see the connection between this job and making a huge difference. How cool would it be if one day I could look back to every project and know I was able to offset my carbon footprint by the places I designed?"

— Christina Guichard, Glumac

INTRODUCTION

"This is your new iPad," announced President Steven Straus during the first-ever Glumac leadership retreat on a sunny June day, 2010, at the Portland Art Museum. Rather than a conventional meeting of the firm's managing principals, he had gathered together thirty employees, most under the age of forty, from all offices to describe and forecast the future of Glumac's place in a rapidly evolving profession. For Straus, Apple's newly-introduced device offered a powerful symbol – of redefining digital work and life and of their own rebranding efforts as a company ("Like the commercials, we want to be the Mac Guy, not the PC") – in their drive to transform the region's built environment. These new tablets, he explained, would be ideal for most communication on the road: reading reports, accessing emails, writing, photography, creating slideshows; likewise, number-crunching tasks would still require server-based, desktop systems for access to Revit and other data-driven platforms. "The lines are blurring, the workplace is evolving," he added, "and I want to make sure we accommodate that new generation of thinkers."

As attendees unboxed and activated their iPads, the stage was set for two days of intense brainstorming on the changing use of technology, improving upon existing processes and collaboration with clients, adding staff expertise, and focusing on new and sometimes unexpected approaches to sustainable design. There would be an element of forecasting here, but also "backcasting" – a technique Glumac engineers often utilize during design charrettes: examining solutions to a project (for example, college buildings) from twenty-five years or more ago allows for closer scrutiny of how much MEP designs have changed; in turn, that may lead to ideas on how processes, tools and design solutions could change in the years ahead. "As you think about the future of Glumac," Straus concluded, "consider this: you are the ones who will be running the company. So what's it going to take to be the best possible firm we can be ten, even twenty, years from now, and how will you help to achieve the goals you set here today?"

"The total radiant heat energy emitted from a surface is proportional to the fourth power of its absolute temperature."

— *Stefan-Boltzmann Equation*

INSTANT, INFORMATION-RICH, IN THE CLOUD

Talk of technology seemed the easiest, most logical place to begin the discussion. Building on the benefits of mobility, Steve Carroll (Irvine) highlighted the coming adoption of paperless processes to manage the large amount of documentation required, for example, in commissioning. "Using an iPad or smart phone allows you to take pictures and automatically create reports, making the process much more efficient," he noted. "Soon, we won't have to rely as much on binders or carrying drawings around, so we can submit our reports instantly, in the field, that day according to the job."

The industry's rapid shift towards the "cloud" – web-based resources and software – emerged as another clear theme. Whether to collaborate with clients more effectively or to access affordable IT services, a growing number of design firms utilize these web applications to create, store and track documents (LEED-Online) or manage green building analysis (Green Building Studio®). Engineers like Skander Spies (Portland) envision new web-based tools to run design simulations more efficiently than the current practice of moving large architectural files back and forth between the engineer's and the architect's servers. "We're going to need to find ways to open up our information services directly with architects," he said. "How great would it be to have access to one model built by the architect and then work off that?"

Continuous refinements in energy modeling, CFD (computational fluid dynamics) modeling, and 2-D software integrated with apps like Google SketchUp – all will play an increasingly critical role in concept design. However, it is the potential for monitoring and tracking building performance over time, and making periodic adjustments, that garnered the most attention:

Georginna Lucas (Seattle): "It would be great if you could go back to old buildings, install a control system to see how it's been working...and determine energy usage."

Brian Berg: (Irvine): "We're actually doing that in our office right now, metering energy, continuously looking at performance, finding out how the lights work...and how to make the space work even better."

Mario LaMorticella (Portland): "We're talking about creating some sort of metering standard, a building dashboard, right now, so the engineer can monitor performance using a website to make sure systems remain efficient from the very beginning."

Mitch Dec (Portland): "I see where you're going with that idea: it would be important to define what type of control points we need to monitor, based on the different types of systems."

Lucas: "In Washington State and Seattle, energy codes now require submetering for bigger buildings in particular: plug loads, the lighting system, definitely various portions of mechanical systems."

David Summers (Los Angeles): "I like the idea a lot. You could develop a standard format for twenty or thirty projects, and as the database grew, we'd have the same data for every project."

TO CREATE LASTING CHANGE

What if energy prices were to double or triple over the next few years? What new markets and/or demand for expertise would that open up within the building industry? Glumac's young leaders were clear and specific with their answers: more CFD modeling, more retro-commissioning, and more renewables – while adding several new disciplines to augment the company's current MEP engineering services.

CFD modeling, while not yet widely embraced by clients and building owners, offers many new opportunities to improve design – glazing selection, placement of wall insulation, airflow, and so on – and, ultimately, building performance. Michele Reesink (Sacramento) observed, "What do we currently have now that can differentiate us from other firms in five or ten years? I see CFD modeling as one powerful solution." The key, she noted, involves expanding its

use within Glumac's toolbox to make the benefits more tangible and demonstrate how it brings value to design decisions for each project. That led to the idea of developing an ROI chart for this technique, to measure its effectiveness and validate the upfront cost much in the same way Glumac currently promotes energy modeling and auditing. Mitch Dec gave a recent example of one university where the engineer disputed its value, claiming it would not work given the thermodynamics and airflows of their systems. Instead Dec noted, "I've offered a small sample of CFD to illustrate the types of answers we can provide. My hope is they'll investigate this further and that more architects start adopting CFD."

Likewise, measurement and verification hold great promise for more building assessment and tracking metrics over time, as more owners request – even require – these steps to ensure their properties are energy efficient and performing optimally. "Do we think it's a good idea? That's where it starts," noted Steve Carroll. "So it's up to us to go on the road, to approach every owner with measurement and verification ideas: this is what you should do, this is why you should do it, and this is how much it's going to save you in operating costs."

The 2008 – 2010 recession clearly became a game-changer for the building industry. Several participants reflected on how the market became more competitive while also redefining many roles and relationships. "We've had to be a lot more creative in selling our

HVAC RELIEF FOR A NEW CORPORATE HEADQUARTERS

1 Green roof
2 Insulation
3 Thermally activated slabs
4 Radiant fin tube heater
5 Passive radiant chilled beams
6 Dynamic shading device

7 Motorized dampers
8 Raised access floor
9 Underfloor displacement
  ventilation system
10 Relief shaft
11 Relief fans

DEEP GREEN CONCEPTS: Glumac's cutting-edge approach to ventilation and thermal control for a new office under development puts the emphasis on architectural integration between the served (occupied) and service spaces — while creating a visual aid to sustainability through use of transparent supply and relief shafts.

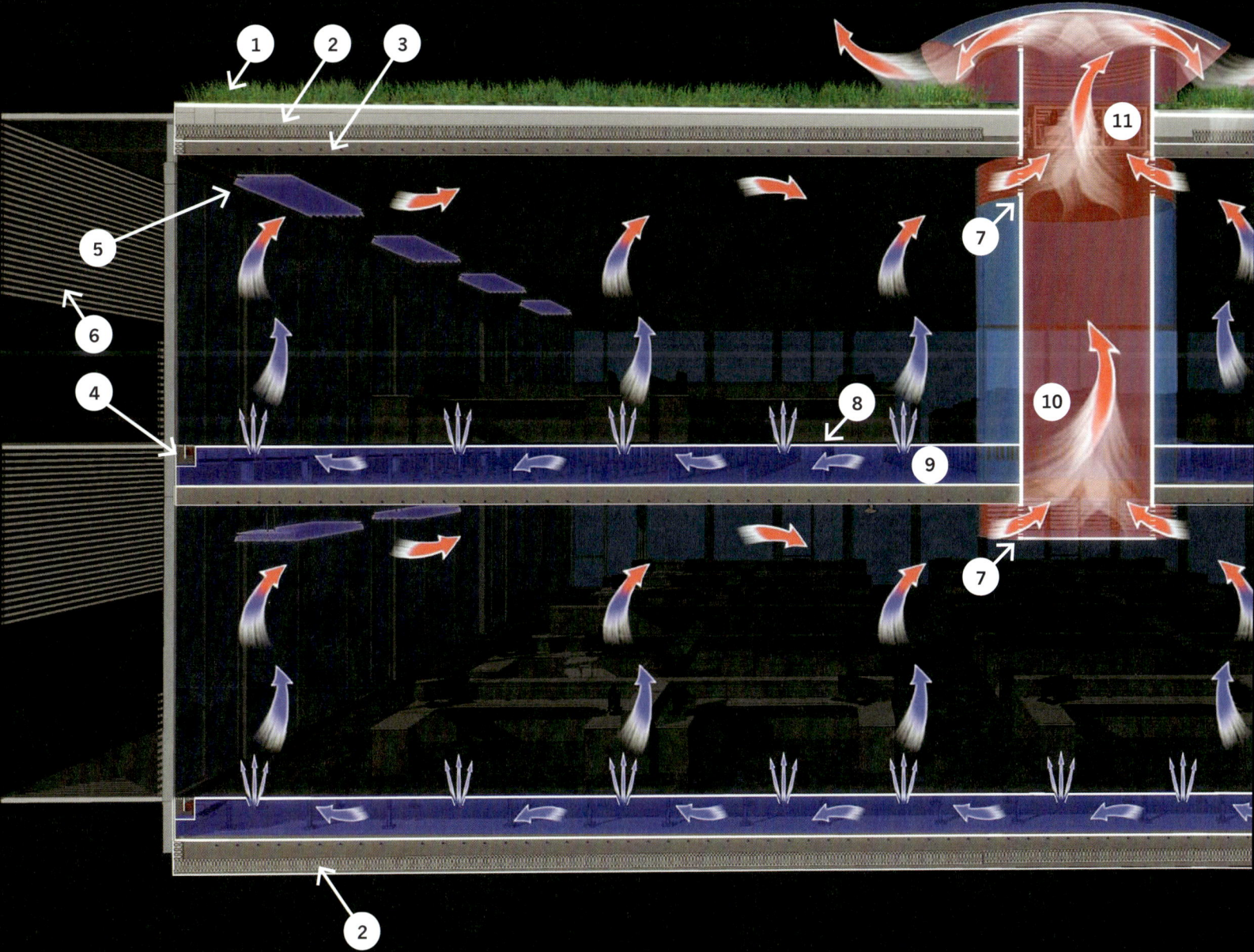

"We can easily say that technology will play a much larger role in what Glumac is going to be than it does today – much more than even ten years ago, or thirty years ago. If you think of that evolution of when engineers were doing their drawings on vellum, or the advent of the first fax machine, the first computer...and what does that look like: two, five, ten years from now? I think that's a question we have to answer."

— *Angela Sheehan, CFO, Glumac*

services," stated Scott Vollmoeller (Seattle), "pushing us to seek more sustainable ideas and measures in which to collaborate with design-build contractors. Our core focus has always been MEP; but now our other services like commissioning, energy modeling and CFD modeling have really come to the forefront."

Angela Sheehan (Portland), Glumac's CFO, recognized even broader implications with that market shift:

> "I think the business community as a whole has changed. And like us, organizations are looking for ways to make a dollar go farther, to reduce their footprint and their overhead, and to make more effective use of the spaces they occupy. That's where we, as a design firm, have a lot of expertise – making good use of smaller spaces – and then helping clients stretch those dollars while still creating effective work environments."

Given the market trend toward renovation and energy upgrades for existing buildings and less focus on new construction, Mitch Dec is seeing an increase in LEED Existing Building Operations and Maintenance projects:

> "A lot of technology is available to improve energy performance in those buildings today. So I see making a 50 percent energy reduction in an existing building as a bigger

deal than designing a brand new building to be net zero. Owners are starting to make investments to attract more tenants by making their buildings well insulated, bringing in more daylight and pursuing other measures. That's going to be 90 percent of the market in a couple of years."

Where modeling and commissioning services serve to expand the company's capabilities, many also anticipate the addition of new skill sets that complement its core engineering expertise. This expansion would include a façade engineer, an elevator engineer, landscape architects to advise on green roofs and storm water management – even adding a behavioral scientist: "someone who understands how you feel in a space based on the light, air quality and temperature." Max Wilson (Seattle) also proposed hiring dedicated programmers: "We're at the forefront of adopting Revit, and yet we haven't fully explored the inner workings, the structure, the programming language, how to customize all the menus, applications and scripts. Utilizing feedback from engineers would allow us to modify its capabilities so the whole company could use this platform to its full potential."

FUTURE PROOFING

Even with a portfolio full of award-winning buildings, Glumac and its engineers look ahead to pushing their designs further – to meeting the 2030 Challenge, net zero energy and water performance, the Living Building Challenge, and beyond. Among the most compelling

HVAC SUPPLY FOR A NEW CORPORATE HEADQUARTERS

1 Green roof
2 Insulation
3 Thermally activated slabs
4 Radiant fin tube heater
5 Passive radiant chilled beams
6 Dynamic shading device
7 Motorized dampers
8 Raised access floor
9 Underfloor displacement ventilation system
10 Presurized plenum shaft
11 Air handler unit
12 Inlet filter screen
13 Pre filter
14 Carbon filter
15 Cooling coil
16 Final filter
17 Supply fans

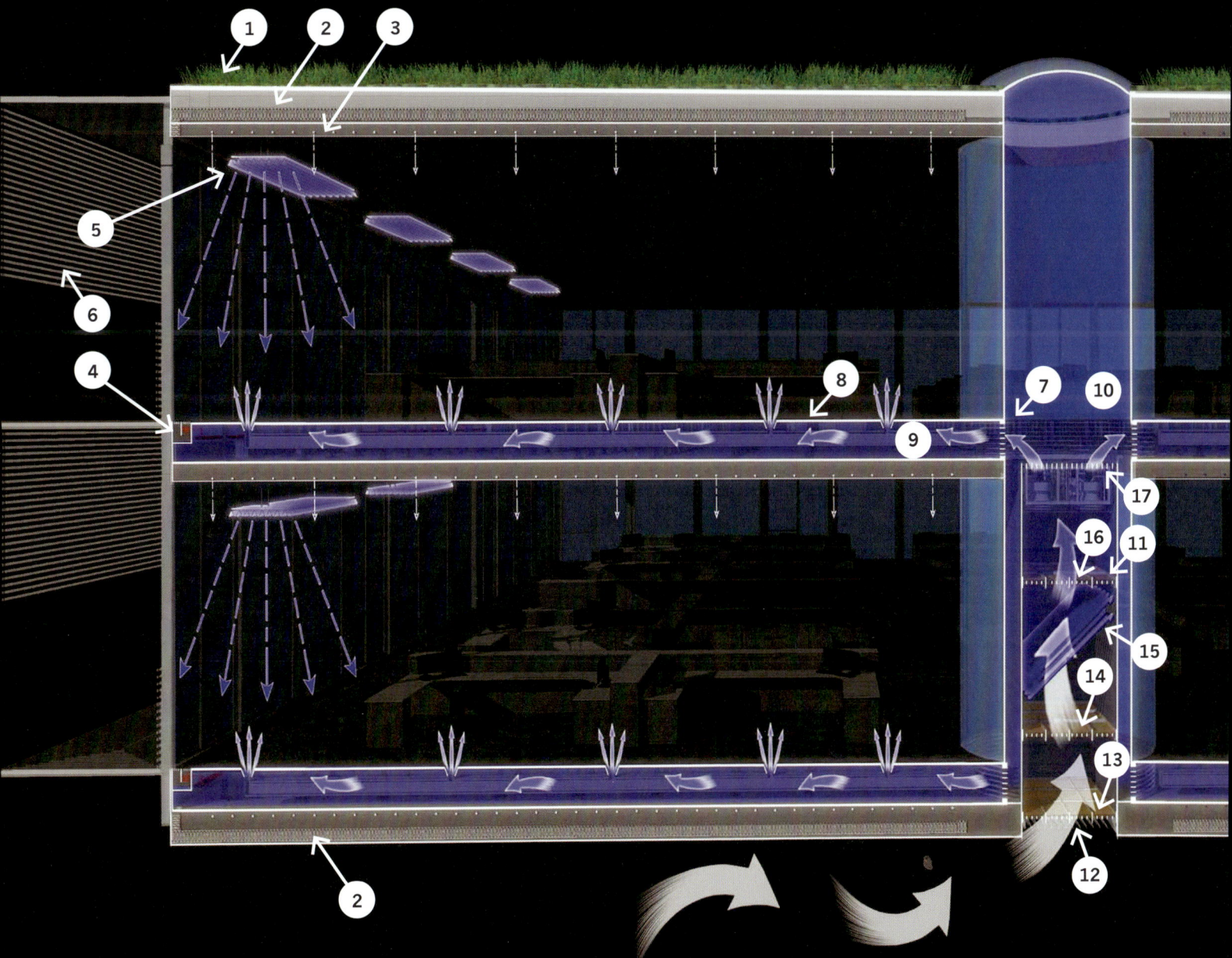

COLLABORATIVE DESIGN: Integral for any
of Glumac's sustainably designed projects,
ongoing collaboration to review and refine specs
has been critical to achieving cost savings and
meeting high-performance building goals.

"Definitely in five years, if not in three years, we're going to need to find ways to open up our information services directly to the architects. How great would it be to have access to one model that's built by the architect and be able to work off that?"

— *Skander Spies, Glumac*

ideas expressed during the leadership retreat, perhaps, was that of "future proofing": in essence, planning a building to accommodate future innovations, particularly where the design concepts or technology may be too costly or difficult to apply in today's market. Mitch Dec noted the importance of making provisions for photovoltaics. "Instead of simply putting off solar today versus ten to fifteen years from now, I'm seeing more discussions around how we can get buildings PV-ready, so they can make these additions later and possibly achieve net zero energy goals."

Glumac designers expect future opportunities with building-integrated wind turbines in urban locales, like the installation on Portland's Twelve|West high-rise (see page 262) and those Steven Straus has proposed for the Embarcadero Center in San Francisco. Fuel cells represent another breakthrough; although the technology is maintenance heavy and still expensive, engineers continue to research future optimal placement (sleeves, conduits) and design applications for this future power source. Future proofing for water means plumbing dual systems in cities: first, as potable lines for drinking water and laboratory lines; and second, the industrial water supply – for irrigation, toilet flushing and other applications.

Upon entering the second decade of the 21st century, *tomorrow* has already arrived in many ways for Glumac, using its own offices to demonstrate the possibilities of sustainable design.

Michele Reesink leads weekly tours of the firm's Folsom, California office (twenty-five miles east of Sacramento) to showcase design elements such as 100 percent daylighting, ice storage, rooftop PV, solar water heating and more as a near-carbon neutral building. Further south, Jennifer Berg (Irvine) takes pride in Glumac's Orange County, California office space as a model for state-of-the-art renovation and market change: "The Irvine Company [the building owner] was initially resistant to what we were trying to do there, explaining the retrofit didn't represent their standard. Now they love it and continually ask to take future prospects through our space as a prime example of green tenant improvement."

Before adding any new expertise or specialized disciplines, before widely adopting cloud-based technologies or tackling its next generation of new buildings, Glumac's path to the future lies *within* – collaborating, continually evolving and improving as a firm, noted Angela Sheehan:

"We have to define as an organization how we would measure success. And my guess is that it would be based upon the individual success of a project, of making sure that we're consistently getting the right people on the right projects. But it's that first big step that matters most – of changing the hearts and mindsets of the people managing the work and the process across the organization to do what's right for the client."

STEVEN STRAUS:

DRIVING CHANGE, A FUTURE OF POSSIBILITIES

Instead of projecting out twenty years or more, you prefer to focus on the importance of fundamental changes over the next five, maybe ten, years as having the biggest impact on engineering and the built environment. How so?

The first big change I see ahead is the idea of plan-and-spec work. That way of doing business is over, and it should be over. It is not a productive, efficient way to build a building. So the negotiated method with general contractors and major subcontractors is the wave of the future. Integrated project delivery starts with the owner, the contractor and design team – all adding to a mutually beneficial relationship. The way to achieve this is by fixing the fees up front: pay this much for overhead conditions and this is your profit, possibly with some bonuses for beating the schedule. And everything is open book, which will encourage more collaborative behavior. You also need to include some verification of the project at some point, so everybody understands that only excellence matters.

Technology is traditionally a key indicator of the future. From BIM and CFD modeling to your recent firm-wide introduction of the iPad, Glumac has certainly gained a reputation for embracing technology, early and often.

And yet, every industry but construction has modernized: the auto industry, agriculture – it's all been mechanized and modernized to focus on efficient production. There's no question that the construction industry is going to change, because if you want to be the best contractor, then you have to figure out how to deliver the product better than it has been delivered in the past. That means utilizing Revit and other software and modeling systems to more effectively automate the building design process.

Steven Straus

Collaboration represents another, still-evolving change in the industry.

Yes, and we need engineers on our team who understand and practice collaborative behavior, who work well with the contractors and other members of an integrated project team. Historically, there have been a lot of problems between contractors and design teams. Sometimes design teams *are* arrogant; and sometimes contractors *do* come to the table with the idea that designers are just theorists, that they don't know how to actually build anything and certainly don't know how to build it cost effectively. Somehow, those barriers have to be broken down.

Selfishly, I think engineers' role in the business is going to become much more integral to the design – certainly more than it was ten years ago. In the past, we weren't consulted much. During the conceptual design, the architect drew the building up and said, "Here it is, go make it work." If we offered input, their response typically was "You can do whatever you want as long as you don't change my design." That's changed dramatically, and now at least the question is being asked: How can we design efficiency into the building? I think you'll see engineers taking a much more proactive role in the design process than they have in the past. Historically, it has been hard to get good engineers: the brilliant "A" students would go to work for Intel or some dot.com and make $150,000, and the "C" students would become an HVAC engineers. Now that's changing, because everybody's really focused on sustainability.

You once said if a prospective client is not interested in sustainability, they should go talk to another firm. So as Glumac tries to push the envelope of MEP design and decides that LEED Platinum doesn't go far enough in this age of rising oil prices, what's next?

Pursuing ever-greener buildings is one area where we can make a difference. But as designers, we also need to consider the embodied energy of building construction. You can make any building carbon neutral by putting enough PVs on it. But if the PVs were made in China where many PVs are produced with coal-fired power, we must consider that those chip plants suck unbelievable amounts of energy to melt silicon, then slice, polish and chemically treat the chips. The process creates an incredible amount of waste. It could take ten years of using photovoltaics just to generate enough power to pay off the energy it took to manufacture.

So, besides creating our carbon calculations to determine how much carbon each building consumes, we need to help architects understand the differences between the embodied energy of steel and concrete and other products. Then we must make a joint decision that results in the lowest carbon footprint overall while understanding what makes a great building: what makes you and I want to sit next to the window? I think there are so few examples of great, sustainable architecture that creates a wonderful environment that also maximizes energy efficiency.

In one possible future for buildings, why couldn't engineers design based on ideas such as biomimicry?

I'm fascinated with biomimicry. We need to study our building elements and create more design approaches based on what occurs in nature. How do animals breathe? How do buildings breathe?

And "future proofing"?

This whole idea of carbon neutral buildings may not be practical to accomplish with today's technology unless it's applied to a small-scale building. So future proofing examines the idea of where building design will be in the future, especially in terms of technology, and what we

ON LEADERSHIP: Steven Straus –"If you're not having a lot of fun at what you're doing, you don't do a very good job. If you're not passionate about the work, if you're not passionate about the client, you end up doing mediocre work."

can do as engineers for little or no money to accommodate that. It's important to plan a building for Day one, but also for twenty, fifty, one hundred years from now.

You also point to creative, "right brain" skills as the ultimate driver of change?

The future calls for making our company, and our people within that company, utilize more right brain skills and abilities. You're a much better person when you have a balance, and it allows you to collaborate with people from all different parts of the business. I think an architect is much more appreciative of someone who goes "Wow, it would be really cool if we could make these lights symmetrical," as opposed to "I'm just an engineer, tell me where to put the lights and I'll wire them for you." Engineering companies continue to focus on and invest in left brain skills: spreadsheets, mapping to specifications, etcetera, instead of inspiring people to build their right brain. So I'm looking to add some really bright people who know more than exactly how light or radiation or even a chilled beam works. There's more to life than just numbers and BTUs and VAV boxes; there's also daylight, there's art, there's music – all of these things that are really spectacular.

FUTURE TECH:

THREE CONCEPTS CHANGING THE WORLD

No matter what form or features Glumac's buildings take in the years ahead, the firm continues to play an active R&D role in the pursuit of new technologies to improve performance within high-rises, healthcare facilities, corporate headquarters, on university campuses and elsewhere.

FUEL CELLS

Even as manufacturers continue the decades-long quest to develop stationary fuel cells for clean, affordable electricity, Glumac wants the State of Oregon to press harder to promote this technology. Currently, the firm is working to identify twenty test sites in Oregon while collaborating with the governor's office to attract fuel cell producers like United Technologies and Bloom Energy to the state. In addition, they are requesting that Oregon's Department of Energy approve tax and energy credits more favorable to fuel cell technology and further encourage its adoption for residential and commercial applications.

Meanwhile, Glumac also continues its participation in research projects such as the E^6/Net Zero Assessment conducted for Portland Community College's (PCC) Sylvania campus. The project team, which includes Gerding Edlen, GBD Architects and Glumac, proposed seven different scenarios to address climate change, environmental stewardship and green workforce development for the institution. Their seventh option, as a net zero campus, would include an innovative waste-to-energy system utilizing a 1.1 MW fuel cell to produce electricity.

PERSONAL SPACE: A core consideration for every Glumac design focuses on controlled environments, allowing occupants to manage their own comfort – as in the perimeter offices at OSU's Kelley Engineering Center.

ACTIVATED FAÇADES

Glumac's design experience at OSU's Kelley Engineering Center, and more recently with ZGF Architects at the Twelve|West high-rise in Portland, has propelled the firm into a multi-year investigation of designing and improving activated façades. Applications include façades that feature motorized operable windows, building-integrated photovoltaics (BIPV) or automated shading devices to manage solar gain at the building's skin while aiding interior heating and cooling systems.

Early in 2010, Glumac sponsored the launch of the Advanced Façades Research Collaborative at the University of Oregon's School of Architecture & Allied Arts in Eugene. This research program includes a firm-taught studio (with a competition element), a research phase (based on winning designs from that studio), and an implementation phase. Its ultimate goal was to design and prototype a model for new activated façade systems capable of retrofitting existing, historic buildings to improve energy efficiency and sustainability.

VACUUM GLAZING

Urban wind turbines, chilled beams, ever-sophisticated building automation systems – all have attracted increased attention as part of a changing palette of sustainable design strategies. Yet Straus and others within the firm see new vacuum glazing for exteriors as having an even greater impact on building performance. With several new products on the horizon, the technology – essentially creating a vacuum between two layers of glass – eliminates conduction and convection while allowing radiation through to better control solar gain within interior spaces. Glumac engineers are now seeking ways to incorporate this alternative on projects, resulting in R-values far above current high-performance glass packages.

PART II

THE PRINCIPLES

PROCESS OF INSIGHTS:
From renewables to radiant
floors, natural ventilation,
building commissioning and
more, Glumac's sustainable
design methodologies adapt
and innovate – maximizing
occupant comfort and energy
performance, often exceeding
project objectives.

CHAPTER FOUR:
THE PRINCIPLES

CALCULATING SUSTAINABILITY: THE MULTIPLIER EFFECT

"Simplicity is about subtracting the obvious, and adding the meaningful."

— *John Maeda*
from The Laws of Simplicity

"It's not enough just to have an energy-efficient building; you also need to create great architecture, a place to be around for the next 100 years or more. And secondly, it should be a space people want to spend time in, a space that's comfortable and serves a purpose, whether as a high-rise office, a hospital, a retirement home or a university lecture hall. For us, simpler is better: to work with architects to design buildings that are naturally lit, ventilated and passively heated and cooled – that's real innovation."

— Steven Straus
President, Glumac

INTRODUCTION

To reveal the true character of a firm requires a look back at its past while peering ahead to its future direction and intent (THE PRACTICE). That understanding grows deeper with evidence of completed works – the mechanical, electrical and plumbing systems carefully conceived and applied to award-winning public and private buildings across a region (THE PROJECTS). Yet, this review tells only part of the story. Equally essential to Glumac's DNA, in fact, are the sustainable design principles that further define the skills, experience and thinking of the firm and its talented designers.

Is it possible to visualize fresh air or capture daylight or save money with rain drops? To measure true comfort for any given space in summer and winter? And to generate and store energy without adding technology?

Every engineer at every Glumac office shares a common *mantra*: "green buildings that work." This philosophy serves as both a premise and a promise that drives the design of every project. It begins with determining the needs and opportunities of the *Building Envelope* and ends with *Commissioning* to verify the performance of every new system or retrofit. In between, Glumac's designers apply their expertise to a wide variety of sustainable design elements, including *Daylighting, Radiant Heating and Cooling, Displacement Ventilation, Natural Ventilation, Geoexchange Systems, Photovoltaics, Solar Thermal, Rainwater Harvesting and Commissioning.*

That process of achieving high-performance, long-lasting buildings also requires a strategic set of tools – tools applied early and appropriately to help inform the client, architect and other team members of the possibilities inherent for every design. Glumac launches many new projects by developing energy pie graphs to depict energy consumption for a code-conforming building, then evaluating each "piece of the pie" for opportunities to reduce usage by at least half over a conventional building. Options may include advanced building façades to save significant energy while increasing thermal comfort and promoting daylight within a space. After reviewing a number of potential energy conservation measures, Glumac runs a series of energy models to demonstrate the benefits of combining sustainable design schemes.

Intuition, context, experience, even experimentation are necessary to advance a new idea. Overlaying every project, Glumac also follows a clear, consistent set of guiding design principles:

» *Design buildings to suit their immediate environment*, rather than attempt to recreate an advanced building design from another locale and a completely different climate.
» *Collaborate with each member of the design and consultant team* as part of an integrated design process to optimize overall performance and cost efficiency.
» *Communicate consistently with all parties.* Particularly, maintain strong interactions and relationships with clients – and make a commitment to stand behind the work.
» *Justify sustainable features by calculating lifecycle costs.* Realizing higher performance, or a higher LEED® certification, may be achievable through smart design without affecting the overall construction budget.

» *Apply expertise from one market area or industry to another.* Project experience in microelectronics, for example, has been used to improve air quality in laboratories – so that new and unexpected ideas may lead to design breakthroughs.

What follows, then, offers a glimpse into the decision-making processes and unique methodologies, tools and techniques that make up Glumac's varied design approaches to energy, MEP systems and more.

FOCUS ON THE
BUILDING ENVELOPE FIRST

The interface between inside and out – what separates the building from unconditioned space: fenestration, doors and skylights, the roof/ceiling assembly, exterior walls, the floor assembly, insulation, and the slab edge.

Creating effective, energy-efficient spaces begins with the building envelope. While many advanced green projects highlight sophisticated mechanical systems and renewable energy applications, they also tend to underutilize passive strategies for heating and cooling through glazing choices, thermal mass and optimal use of insulation. Close attention to the building envelope can reduce the size of mechanical and electrical systems and, in turn, significantly reduce first costs. Likewise, decreasing equipment as well as pump and piping sizes can lead to greater savings while offsetting the cost of higher insulation values, making a project's payback more immediate.

OPTIMIZED GLAZING

Selection of glass, as well as size and placement of windows, become critical considerations for the building envelope in optimizing energy efficiency and thermal performance. Typically, buildings lose eight to ten times more energy through a window (U factor of $0.45/2.55W/m^2 \cdot {}^{\circ}C$) than all wall faces (U factor of $0.06/0.34W/m^2 \cdot {}^{\circ}C$) combined.

Livability of the interior environment, in terms of daylight and connectivity to the outdoors, must be considered as well. Many states now cap the window-to-wall ratio at 40 percent (Oregon's code tops out at 30 percent) for prescriptive code compliance. Glumac designers aim to utilize "whole building modeling" to demonstrate tuned glazing systems that optimize passive heating and cooling and daylight harvesting – often applying 50 to 60 percent glazing on south and north façades and 40 to 50 percent on the east and west façades.

Another important consideration in glazing choice is to determine the optimal shading coefficient for each building's orientation through modeling. A shading coefficient of 0.8 allows a significant amount of solar radiation into a space, while a coefficient of 0.1 reflects the majority of the solar heat away from the building. Clear glass provides effective passive heating but increases cooling energy consumption; inversely, highly-reflective glazing reduces cooling consumption and passive solar gain. Identifying the optimal shading coefficient for each building façade ideally balances heating and cooling energy according to a site's geographic location and the mechanical and lighting systems in use.

PASSIVE POWER: Glumac's comprehensive MEP strategies for a new LEED Gold science building at Azusa Pacific University (Azusa, California) focused on a highly-optimized façade that features multi-colored cement board and full-height translucent channel glass. Combined with oversized air handlers and high-efficiency boilers and lighting systems, performance of the Segerstrom Science Center exceeds California's Title 24 Energy Code by 11 percent.

INSULATION VALUES

Insulation choices also impact the building environment. If the façade is poorly insulated, interiors may require higher winter temperatures and cooler summer temperatures to offset additional gains or losses while achieving the same level of thermal comfort. Glumac designers typically advocate for increased insulation, above what the local energy codes require. Energy modeling can illustrate savings for an additional one to two inches (2.54 to 5.08 cm) of insulation. Load calculations then determine further load reductions and the savings associated with small HVAC equipment.

Traditional exterior wall designs for commercial buildings utilize batt insulation between metal wall studs. Thermal bridging at the metal wall studs degrades performance of the wall R-value by as much as 50 percent. Glumac recommends providing continuous rigid insulation on the outside of a stud wall. In cold climates, the insulation between the studs might need to be eliminated to mitigate the potential for the dewpoint to fall between the two insulation interfaces and condense, creating the potential for mold.

PROCESS/TOOLS

Each Glumac team initiates pre-design and design work with a careful examination of state energy code requirements. While specifics vary, most states adhere to a local energy code and its commercial envelope provisions as defined by climate zone and type of construction. Glumac engineers also consider a building's planned use and programming. As one example, office spaces may be optimally designed with exterior shades and a dual-pass system – utilizing 100 percent outside air for ventilation, with a radiant floor supplying all heating and cooling. Next, the team calculates anticipated energy use and building loads. An energy analyst examines potential energy savings, and the project's mechanical engineer sizes equipment, with an eye toward reducing capital costs and potentially eliminating whole systems (primarily heating) through proper selection and placement of insulation. And finally, designers create Autodesk Revit® models of the building envelope to import into other analytical programs, making it possible to quickly model envelope alternatives during a project's conceptual phase.

FURTHER DESIGN FACTORS

The building envelope also creates important opportunities for daylight harvesting, a strategy to be approached with caution. With the popularity of floor-to-ceiling vision glass (often to increase lease rates) in office buildings, residential towers and lobby spaces, the glass allocated for natural light often exceeds the amount needed. As a result, tenants may close blinds and sun shades, which then disables dimming controls for lighting

AZUSA PACIFIC UNIVERSITY: Completed in 2009, the 72,000-square-foot building wraps around a steel moment structural system – its exterior integrating a "rain-screen" cement board with channel glass, which adds natural illumination to laboratories and classrooms.

WELCOMING GLASS: Designed by Ingenhoven
Architects, the award-winning European Investment
Bank in Luxembourg features 13,000 square meters
of glass to maximize daylighting from all angles.
Walls, floors, ceilings and all other interior surfaces
also contribute to high occupant comfort.

EDITH GREEN/WENDELL WYATT: Modernization of Portland's Edith Green/Wendell Wyatt Federal Building (top and bottom) incorporates a 179kW solar array. The project is also designed to use 40% less lighting energy than Oregon Code. Slated for completion in 2013 and targeting LEED Platinum certification, the project received the first-ever "Portland 2030 Challenge Design Award."

and renders façades with daylighting ineffective. Some buildings use excessive daylighting to save money by reducing electric lighting costs; again, this design may dramatically increase cooling energy demand far beyond any potential savings.

In addition, it is especially important to collaborate with each project's architectural designer on a building façade that incorporates natural ventilation, a sustainable design feature applicable in most parts of the country. Temperate spring and fall seasons offer many hours to take advantage of fresh air, even in regions with extremely hot summers and cold winters.

OPTIMIZE **DAYLIGHTING** OPPORTUNITIES FOR EVERY BUILDING

Using indirect natural light to illuminate indoor spaces as an alternative to electric lighting – while maintaining uniform levels, controlling glare and reflections to create visual comfort and reduce energy costs.

Daylight: bright, plentiful and, when controlled properly, better for many indoor tasks. Daylighting presents a significant opportunity for energy savings, as well as productivity improvements – creating actively-used spaces in all building types. While continually adding to their lighting expertise over the last decade, Glumac designers concentrate on the visual comfort of spaces, striving to achieve high-quality daylighting that actually outperforms electric lighting. Optimized daylighting requires a high level of client commitment, from design (models, mockups, selection of shading devices and controls) through purchase and construction. Ultimately, this strategy is a matter of economics through early decisions that involve the building footprint (width, height, depth), key elements (atriums and skylights), and orientation (whether to include west-facing windows, etcetera). To effectively pursue daylighting, therefore, the client and project team must also evaluate potentially higher first-cost features such as sawtooth roof design or external light shelves, all integrated within the lighting scheme to create an optimal building envelope.

Close collaboration with the architectural team becomes essential as well. In turn, this coactivity leads to better-informed decisions about the depth of bays, interior finishes, window orientation and window treatments – and further ensures that glazing and skylight areas provide adequate daylight without creating glare. Glumac believes most spaces constructed today include more glass than necessary. Daylighting must incorporate shading strategies to accommodate the optimized views and architecture of contemporary construction. Consequently, daylighting design should control solar incidence, so reductions in electric lighting loads counterbalance any increases in heating and cooling energy from windows in the building.

MORE THAN JUST POKING HOLES IN THE ENVELOPE

Even before determining the building footprint or envelope concepts, effective daylighting begins with orientation. In many cases, an optimal floor plan runs east to west. The

COURT RULES: Daylight blends seamlessly with a high-efficiency fluorescent lighting system within the new Health and Wellness Center at Western Oregon University (in Monmouth). Opened in January 2011, the LEED Gold facility added nearly 80,000 square feet of academic, recreation and athletic space to the "Old P.E." building on campus.

northern façade (for the northern hemisphere) receives even light throughout the day and throughout the seasons. When taming direct sunlight, spaces adjacent to the southern façade function well for detail-oriented activities and fabrication processes requiring visual acuity. Eastern and western façades remain the hardest to control, as the sun rises and sets at a very low angle. Glumac's designers always give special attention to glare control. If tinted or reflective glazing becomes necessary as a last resort to remedy glare issues, Glumac recommends the use of spectrally-neutral glass, which does not "discolor" outward views. Additionally, lighter interior finishes, rather than dark colors or wood, aid in managing brightness; for example, bouncing daylight off matte-white ceilings helps to maximize energy savings while creating more comfortable working conditions for occupants.

Daylighting may be incorporated into virtually any structure. As transitional spaces, atriums can afford a direct view of the sky while admitting beams of sunlight, shadow patterns and variable light levels throughout the day. Warehouses function well with little or no directional lighting and allow daylighting via translucent acrylic panels and similar roof materials. Offices and classrooms present more challenging applications, as users occupy them for long periods of time and require even greater visual comfort. Ideally, neither space should allow direct penetration of the sun. This design feature creates better conditions for occupants to concentrate on tasks such as reading, writing or working at computers.

MAKING SMART USE OF LIGHT

The success of daylighting schemes also relies on creative use of overhangs, light shelves, manual or automated louvers, and even strategically-placed vegetation to control light and reduce solar loads – particularly on southern façades. Many Glumac projects include interior light shelves and exterior sun shades. External building elements provide space for PV canopies; however, they may also require added maintenance due to snow and roosting birds, and pose an obstacle for window cleaners. Internal light shelves help to minimize glare in spaces adjacent to windows, yet they can lead to problems with air flow at the perimeter and inadequate heating or cooling scenarios.

Integral for decision-making, Glumac employs CFD models to understand air flows and further accommodate these elements. Designers also recommend operable louvers or shades for clear glazed areas below each light shelf – at times even above the light shelf – to darken the room if needed. A clear indication of failed daylighting, they caution, occurs when tenants use cardboard or other quick fixes to block the glare from windows, which eliminates daylight harvesting capabilities.

PROCESS/TOOLS

Fundamentally, Glumac seeks to understand how light will perform through a series of daylighting studies: modeling its movement and intensity within a space while considering climate

WASHINGTON STATE UNIVERSITY RECREATION CENTER

1 Direct sun light
2 Translucent light diffusing skylight
3 Internal light shelf/ translucent light diffuser
4 Clear angled glass to reduce glare
5 Translucent light diffusing glass
6 Exterior shading device
7 Clear glass
8 Fluorescent lights with dimmer ballast linked to photo sensor

SUNLIGHT POOL: A highlight of the Washington State University Student Recreation Center (in Pullman), the natatorium features a five-lane, 25-yard lap pool and adjoining leisure pool. Glumac's design goal: to provide as much daylighting as possible while avoiding direct sunlight that can cause glare, harsh or veiling reflections, and high contrast within the space – essential in aquatic settings for both comfort and life safety issues. Elements include skylights, translucent light-diffusing glass, light diffusing light shelves, angled glass and exterior shading devices.

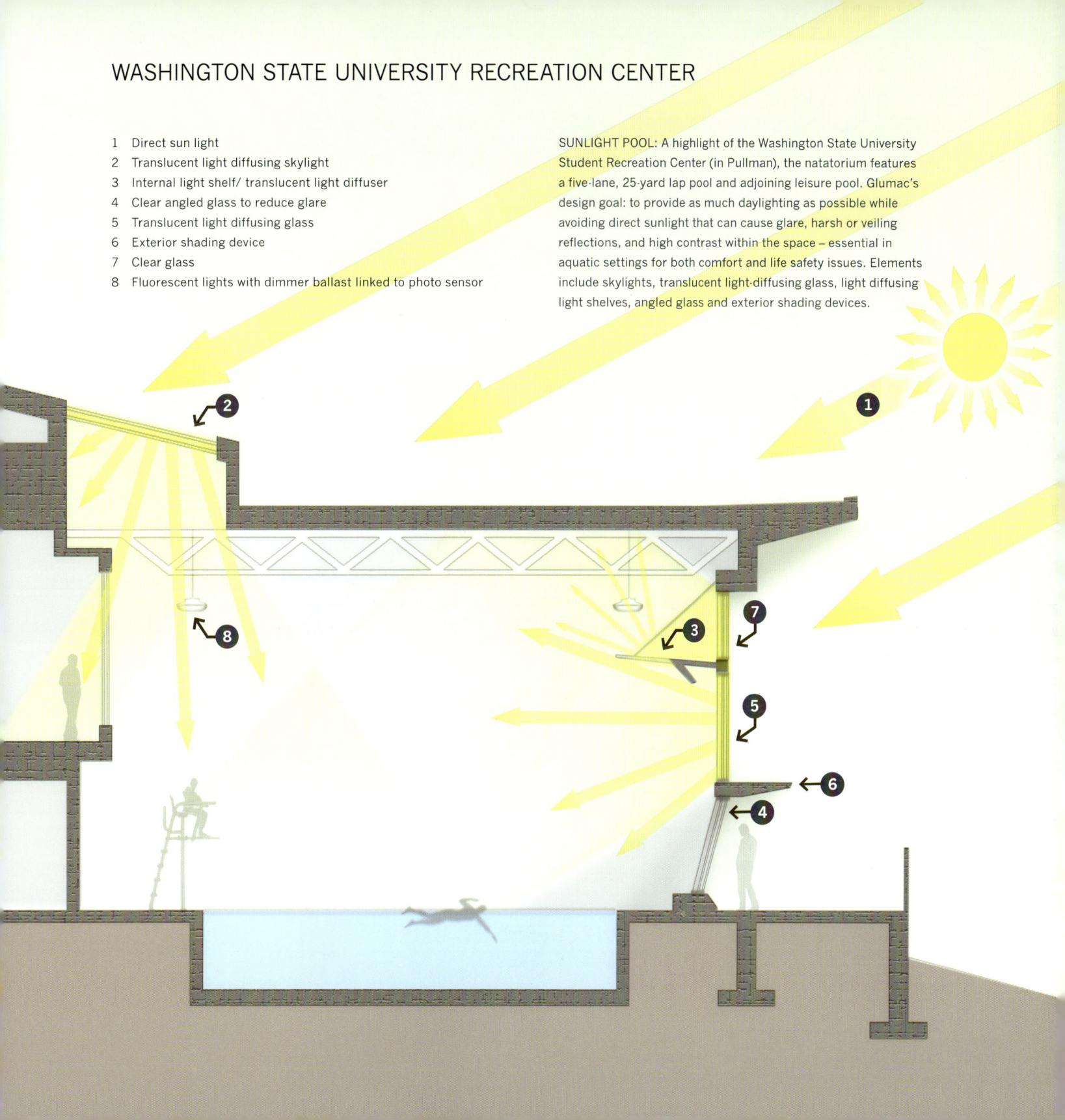

WESTERN OREGON: The University's Health and Wellness Center design integrates daylighting and a highly efficient building envelope, with a combination of displacement ventilation and natural ventilation, to yield 20 percent better energy performance than code.

Summary sun

Winter sun

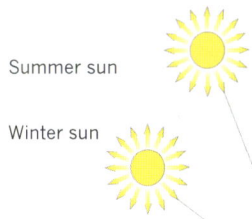

DAYLIGHTING TAMED: An important design detail, light shelves allow for a smooth transition of luminance near windows while providing visual comfort even on sunny days. Electric lights around the perimeter can remain off, and blinds stay open.

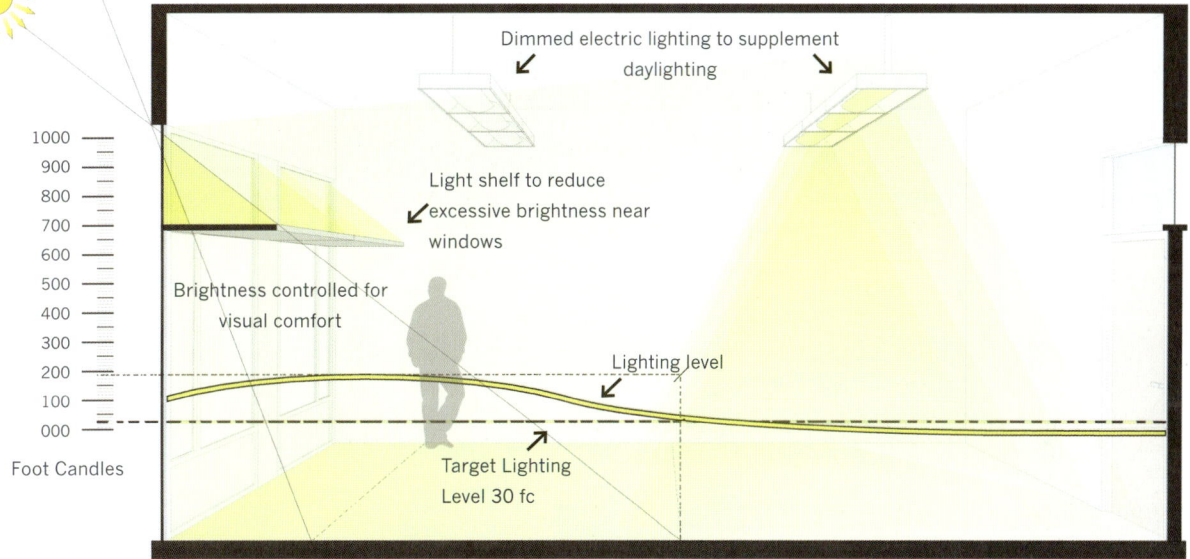

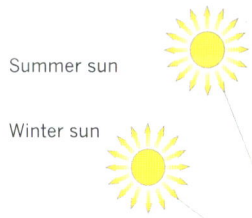

Dimmed electric lighting to supplement daylighting

Light shelf to reduce excessive brightness near windows

Brightness controlled for visual comfort

Lighting level

1000
900
800
700
600
500
400
300
200
100
000

Foot Candles

Target Lighting Level 30 fc

Summer sun

Winter sun

DAYLIGHTING UNTAMED: Without light shelves, the amount of light near windows (inside a south-facing façade) on a clear, sunny day becomes excessive and uncomfortable, causing occupants to pull down blinds and turn on electric lights.

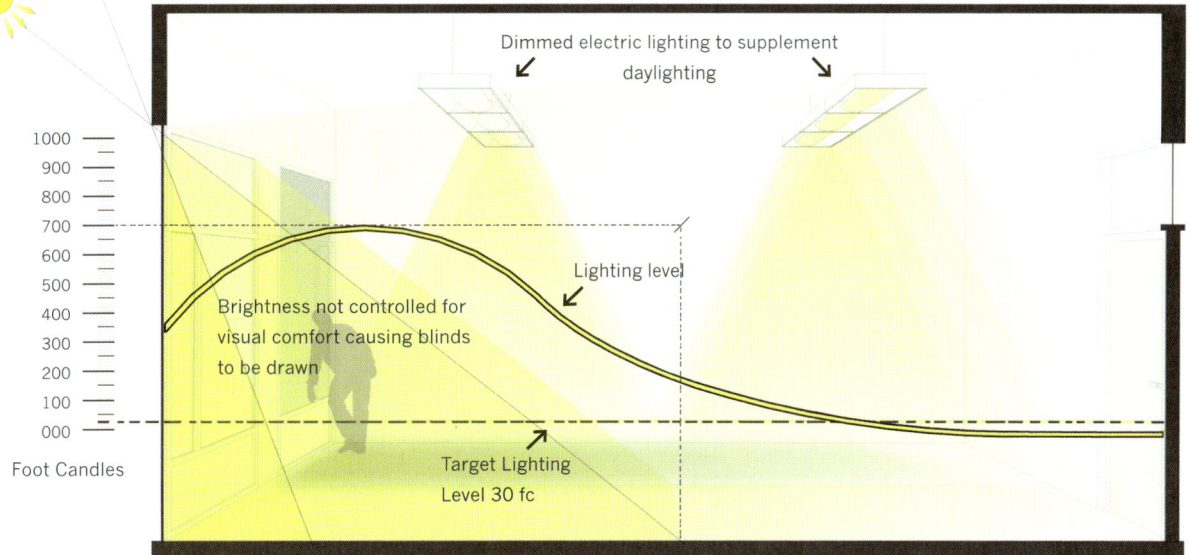

Dimmed electric lighting to supplement daylighting

Lighting level

Brightness not controlled for visual comfort causing blinds to be drawn

1000
900
800
700
600
500
400
300
200
100
000

Foot Candles

Target Lighting Level 30 fc

zones, time of day, cloud cover, season and other variables. Various software packages permit rapid analysis of 3D models which, compared to earlier physical models of cardboard or wood, may be edited quickly as the building design evolves.

Engineers also focus on several key design points to optimize visual comfort for all spaces:

» Keep building dimensions shallow enough to ensure effective penetration of daylight – particularly when not incorporating skylights
» Top-lighting can be more effective than sidelighting
» Design electric lighting for nighttime conditions
» In a well-designed building, windows, skylights and control systems can make electric lighting loads redundant 75 percent of the time
» Sidelighting a building does not require floor-to-ceiling glass; in fact, fenestration only above waist height and a 50 percent window-to-wall ratio can provide sufficient daylighting for most spaces
» To effectively top-light the highest floor of any structure, skylights may be sited on just 5 percent of the building's roof area (as a rule of thumb); larger skylights can be effective with shading devices

FURTHER DESIGN FACTORS

Direct/indirect electric lighting systems play a proven role in daylighting design as well. These luminaires create visual comfort with fewer foot-candles and superior energy savings. Even in the presence of abundant daylight, some occupants prefer electric lights to stay on, dimmed at minimum levels, providing psychological reassurance that all in the building remains functional.

Also integral for daylight harvesting: good lighting controls, particularly daylighting sensors (photocells) located strategically throughout the building while factoring in window placement, window coverings, depth of bays and orientation. Sensor placement and calibration becomes critical, since the color of finishes and proximity to lighting fixtures may produce false readings and oscillating light levels. Finally, new technologies incorporate wireless controls, allowing owners and tenants to relocate sensors as needed after occupancy to further optimize their location.

HEALTH AND WELLNESS: Western Oregon's new academic space uses extensive daylighting and automatic window shading to maximize solar heat gain and increase user comfort.

PERFECT WORKOUT: The Center's recreation component includes an elevated track in addition to the two-court gymnasium, racquetball courts, multipurpose rooms, strength and weight training area, new locker rooms, and a 40-foot-high rock climbing wall.

SUNLIGHTING

CHANNELING SUNLIGHT INTO HIGH-RISE
BUILDING INTERIORS FROM THE WALL

Glumac continues to work with a Canadian company on an innovative technique to harvest daylight from high-rise building facades and then channel the captured light deep into office interiors. The Core Sunlighting System from SunCentral enables electric lights to be shut off up to 75 percent of the time while reducing HVAC cooling loads by up to 25 percent.

This new technology concentrates sunlight tenfold; and because it uses daylight for illumination as is, the system does not suffer from energy conversion losses commonly associated with PV technology. Elements of the SunCentral system include:

» Sealed and PV-powered Sunlight Concentrators, ranging from four to six inches (10.16 to 15.24 cm) in depth, which are integrated into the spandrel areas of curtain wall systems in new building construction projects - or installed on building exteriors in retrofits.

» On the interior, Hybrid Light Guides with wireless sensors and dimmable controls connect to the Sunlight Concentrators. Hybrid Light Guides have artificial lights automatically compensating with electric lighting during cloud cover and evenings. Hybrid Light Guides also feature light fixtures that can be flush or suspended from the ceiling.

The manufacturer estimates commercial payback for the Core Sunlighting System at three to ten years, independent of government subsidies and dependent on the price of electricity and amount of daylight available. Its research efforts are currently funded by non-diluting grants led by the Government of Canada and leading private sector companies

including: Sustainable Development Technology Canada, Natural Resources and Public Works Canada, BC's Innovative Clean Energy Fund, BC Hydro, Philips, Busby Perkins+Will, Morrison Hershfield and The University of British Columbia (UBC).

In addition, six technology demonstrations of the system on actual buildings are currently underway prior to their commercial launch in 2012. In each case, electric utilities will provide independent third-party measurement and verification analyses. Three of six installations have been completed at BCIT, UBC and at a Living Building Challenge Project at Okanagan College – all in British Columbia.

SUNCENTRAL'S CORE SUNLIGHTING SYSTEM:
Inside the building during the evening, electric lights provide illumination. Light fixtures can be flush or suspended from the ceiling.

SUNCENTRAL'S CORE SUNLIGHTING SYSTEM: DAY

1   Ground floor: flush sunlight concentrators and
    sunshade sunlight concentrators
2   2nd floor: 10° angled sunlight concentrators
3   3rd floor: flush sunlight concentrators – integrated
    within the curtain wall
4   Hybrid light guide with integrated light sensors
    and dimming control

CHANNELING LIGHT: Sealed, PV-powered Sunlight Concentrators collect
and concentrate (ten-fold) daylight, which is then supplied to Hybrid Light
Guides and routed horizontally into each floor of a high-rise building.

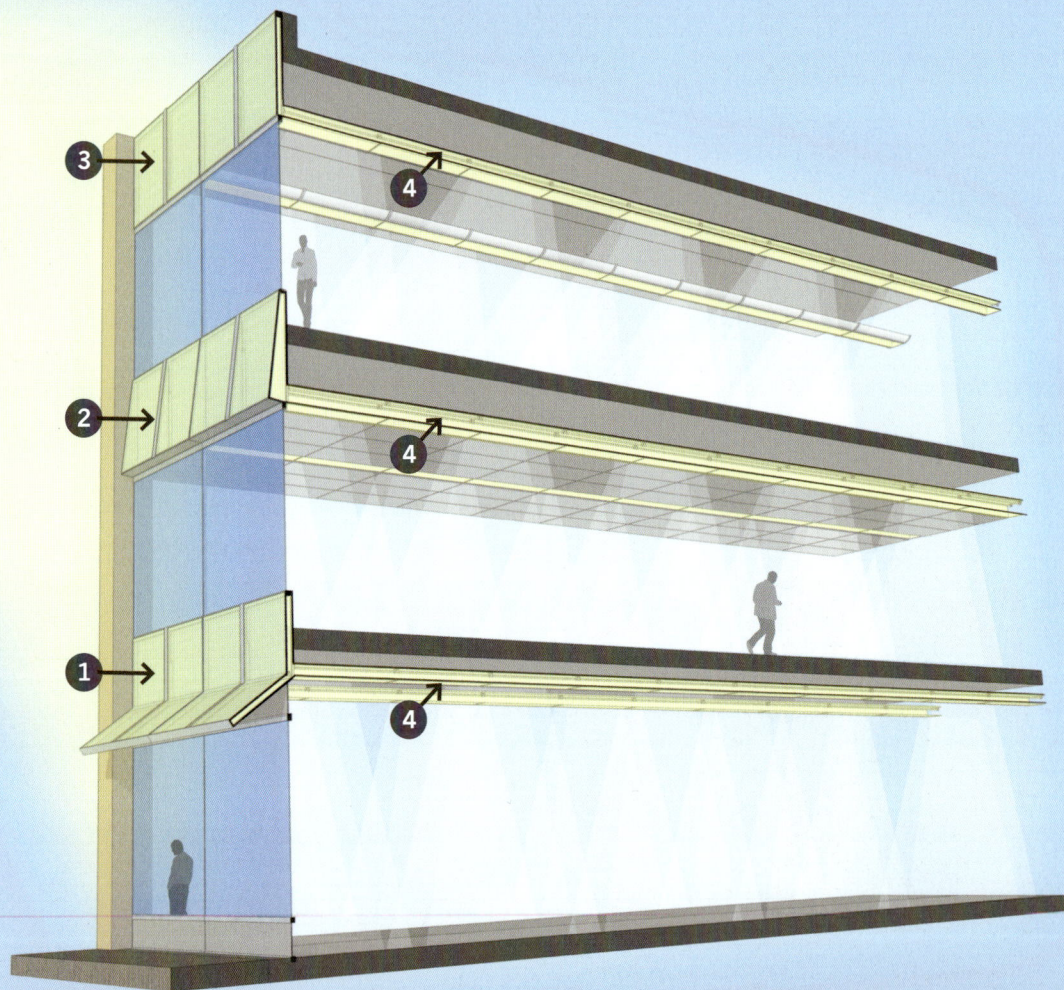

SUNCENTRAL'S CORE SUNLIGHTING SYSTEM: NIGHT

1 Ground floor: flush sunlight concentrators and
   sunshade sunlight concentrators
2 2nd floor: 10° angled sunlight concentrators
3 3rd floor: flush sunlight concentrators – integrated
   within the curtain wall
4 Hybrid light guide with integrated light sensors
   and dimming control

CAPTURED LIGHT: In night mode, Hybrid Light Guides automatically compensate with artificial lights – so the building requires only one lighting system in the evening and during periods of cloud cover.

CONSIDER THE **RADIANT HEATING AND COOLING** ALTERNATIVE

Circulating heated or cooled water through existing surfaces and structures (floors/walls/ceilings) to absorb or release radiant energy to condition a space and to raise or lower air temperatures.

INTEGRATED COMFORT: At Portland's Twelve|West and elsewhere, Glumac designers increasingly specify chilled beams as an overhead radiant strategy for high-rise offices in combination with dedicated outdoor ventilation systems – delivering air directly to the space or via under-floor air delivery systems.

Although an old technology, radiant heating and cooling presents an increasingly popular choice within many modern building designs. Where 100 percent air systems utilize convection primarily to heat or cool a building, radiant technologies condition spaces through a combination of radiation and convection. Radiant systems can significantly reduce energy consumption while providing superior comfort for occupants. Applications include in-slab radiant floors and ceilings (activated thermal mass) and overhead chilled elements. Ideal scenarios for radiant floor cooling range widely – from any indoor space with controlled conditions to those with a connection to outdoors (in climates with low summer humidity) such as entrance lobbies. Active or passive chilled beams also offer an efficient cooling alternative, particularly for cooling-dominated spaces, programs that require room-by-room control, or building operations where it is feasible to decouple cooling and heating loads at a zone level from the central system.

RADIANT FLOORS

In commercial construction, the composition of in-slab systems usually includes HDPE tubing cast in concrete. Water flowing through the piping becomes the transport medium to heat or cool the floor surface and therefore condition the space above. In heating mode, radiant floor slab systems increase the mean radiant temperatures of served spaces; as a result, space air temperatures can be set as low as 65°F (18.3°C) and still maintain excellent comfort during heating seasons. Although less common, radiant floors also provide cooling quite successfully. Tempered water that is 60°F (15.6°C) circulates through the floor slab, maintained at a surface temperature slightly higher than the room dewpoint. This temperature becomes the limiting factor on cooling capacity — important in preventing condensation on the floor.

In both heating and cooling modes, radiant floors consume less energy, requiring about one-seventh the amount of transport energy to pump water versus moving air with a forced air system. While conventional systems utilize 48°F/8.9°C for cooling and 140°F/60.0°C for heating, radiant systems narrow that range to 60°F/15.6°C water for cooling and 90°F/32.2°C for heating. This capability facilitates the use of earth coupling, super-efficient condensing boilers, nighttime evaporative cooling with cool water storage, and solar supplementation.

OVERHEAD RADIANT SYSTEMS

Overhead radiant systems – integrated into ceilings or suspended below the ceiling plane – provide surface area for both radiant and convective conditioning. The amount of space required will depends on the type of equipment. Occupants also feel the radiant effects, contributing to their comfort.

According to rated capacity per area, overhead radiant falls into four categories: flat panels, chilled sails, passive chilled beams and active chilled beams. Passive chilled beams deliver significantly more capacity per unit area than flat radiant elements and function much like a radiator: circulating tempered chilled water to produce both a radiant and convective cooling effect. As air comes in contact with the chilled beam, it becomes more dense and falls through the beam, inducing air flow through the coil. Glumac designers use passive models strictly for cooling. Active chilled beams provide the highest capacity per unit area. These devices cool entirely by forced and induced convection, with the chilled beam connected to the ventilation air system and supplying air directly into the space. Similar in nature to induction units along the perimeter of many high-rise office buildings of the 1970s, this technology fell out of favor because the equipment took up floor space. Today's active chilled beam, however, achieves comparable energy efficiency and comfort levels without reducing floor area.

PROCESS/TOOLS

Glumac's design process begins with a finite element analysis to determine the cooling capacity of a radiant floor system. Because humidity control tends to be difficult in entrance lobbies and similar spaces, engineers work to ensure the supply air's dewpoint falls below the radiant element surface temperature to avoid condensation. Capacity of the radiant element then becomes a function of a surface temperature above this dewpoint. Determining surface temperature can inform load calculations and control sequences to ensure a floor system is operating within design parameters.

One challenge in configuring overhead radiant systems involves economically defining the number, size and placement of radiant elements necessary to meet the cooling capacity and comfort needs of a building space. Flat panel sizes in a passive radiant system, for example, run the risk of being so large they dominate the ceiling plan. Similarly, while higher capacity chilled beams require less area, their high unit cost requires careful consideration related to budget. To optimize a radiant system, thorough analysis may include both CFD modeling and comfort analysis using a tool like the Center for Built Environment's Advanced Human Thermal Comfort Model.[1]

1. This computer model of the human body is sensitive to detailed thermal complexities around the body – and is capable of modeling the indoor environment in detail to predict comfort and "thermal perception."

LINEAR ACTIVE CHILLED BEAM AIR CIRCULATION

1 Linear active chilled beam
2 Fresh air supply duct
3 Flex connection
4 Volume damper
5 Cooling coil

6 Warm air recirculated through cooling coil
7 Recirculated air mixed with fresh outside air and supplied to room
8 Suspended ceiling

LOW ENERGY, HIGHER OUTPUT: Integrated with a building's primary ventilation air supply, active chilled beams deliver higher air velocities and more cooling potential (about two times higher) than passive systems. Primary air passes through an induction nozzle, inducing additional airflow from room air (secondary air) through the cooling coil and down to the conditioned space.

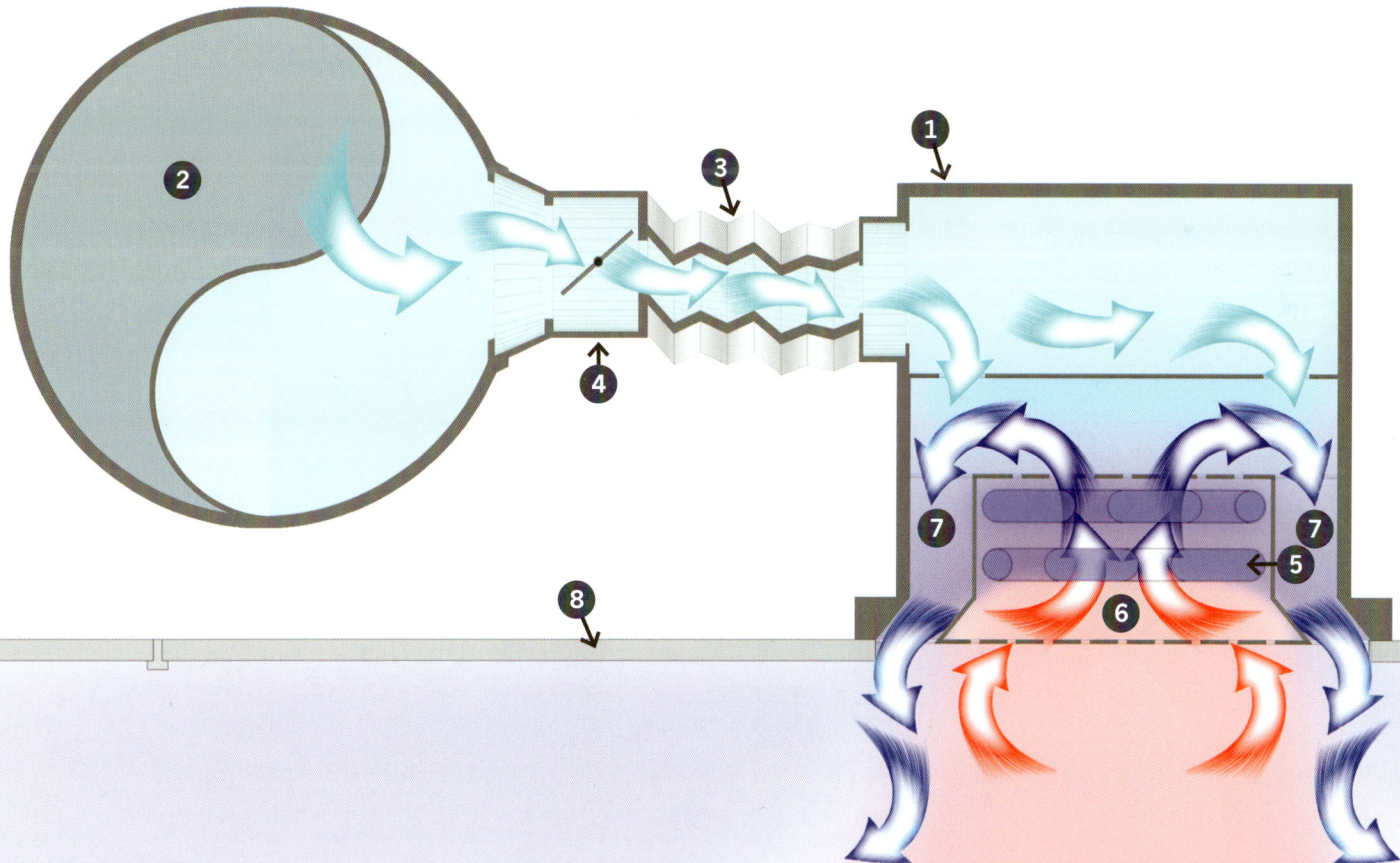

525 SOUTH FLOWER STREET RADIANT HEATING AND COOLING

1 Passive chilled sails
2 Chilled ceiling
3 Active chilled beams
4 Underfloor air distribution in open office area
5 Perimeter convector
6 Atrium with stack ventilation
7 Skylight with motorized windows for natural ventilation

HIGH DENSITY COOLING: To meet financial goals for the office renovation, Gensler-Los Angeles increased its space density with a mezzanine level for added staff.
As a result, Glumac recommended chilled sails to accommodate this low height – supplemented by ventilation air through the underfloor displacement ventilation system.

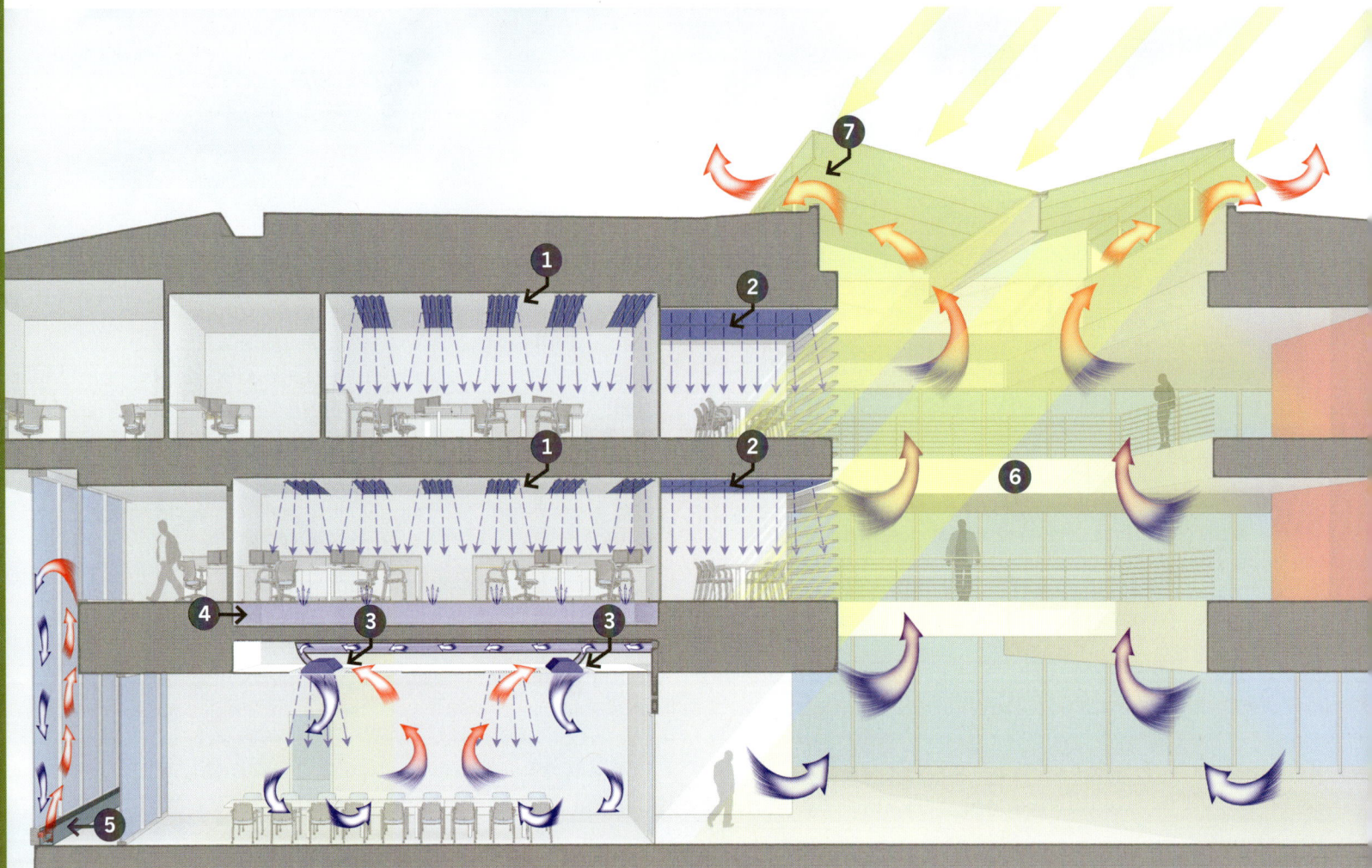

CHILLED SAILS: Gensler's new downtown
Los Angeles office – designed for LEED
Platinum certification – demonstrates
how chilled sails, chilled beams and other
radiant technologies can be incorporated
successfully into new construction or
building retrofits.

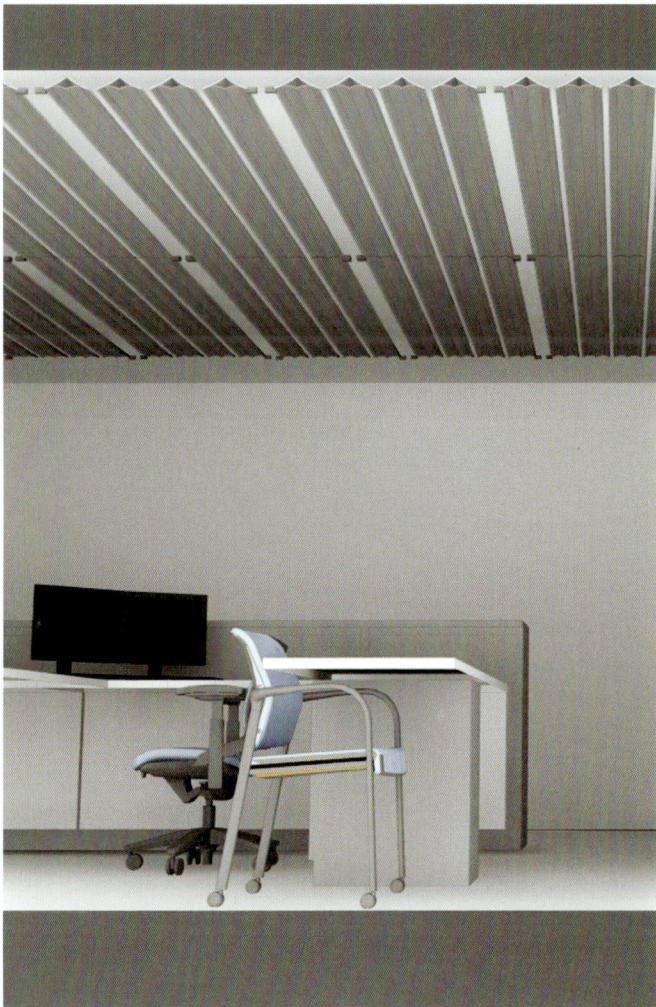

FURTHER DESIGN FACTORS

All radiant installations require a ventilation system as a source of outside air for occupants. In winter, a radiant floor can provide 100 percent of the heating capacity for most spaces; during warmer weather, it may support approximately two-thirds of space cooling, with the remaining one-third handled by the air system.

Overhead radiant systems also have limits: maximum cooling capacity varies from approximately 10 to 250 BTU per hour per square foot ($32W/m^2$ to $2,700W/m^2$) of installed device. For low density devices, in particular, perimeter loads on a building's south and west sides often exceed capacity, making supplementary cooling necessary. However, where designs limit glazing to 30 percent of the building façade – and by incorporating shading devices – radiant ceilings may provide adequate cooling for both perimeter and interior spaces. Evaluating whether to install active or passive chilled beams comes down to cost, energy efficiency, cooling capacity and the aesthetic goals of a space. Passive systems perform well for applications requiring very low loads per square foot/meter and along perimeter spaces with a high performance envelope or glazing, well-shaded exposures, etcetera. Alternatively, active systems deliver a higher volume of air and more cooling capacity while utilizing a smaller panel area.

PASSIVE CHILLED CEILING PANEL SYSTEM

OVERHEAD RADIANT: High capacity, quiet and compact, these simple systems pass chilled water through copper piping; made of highly-conductive metal, the ceiling panel disperses the energy, then uses radiant cooling to provide comfort for the occupants.

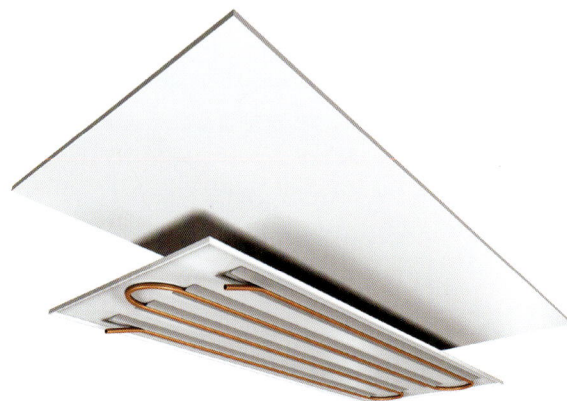

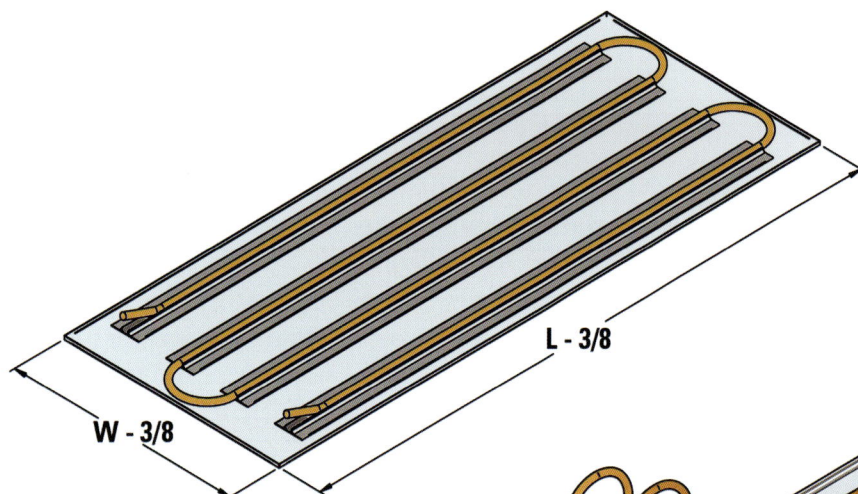

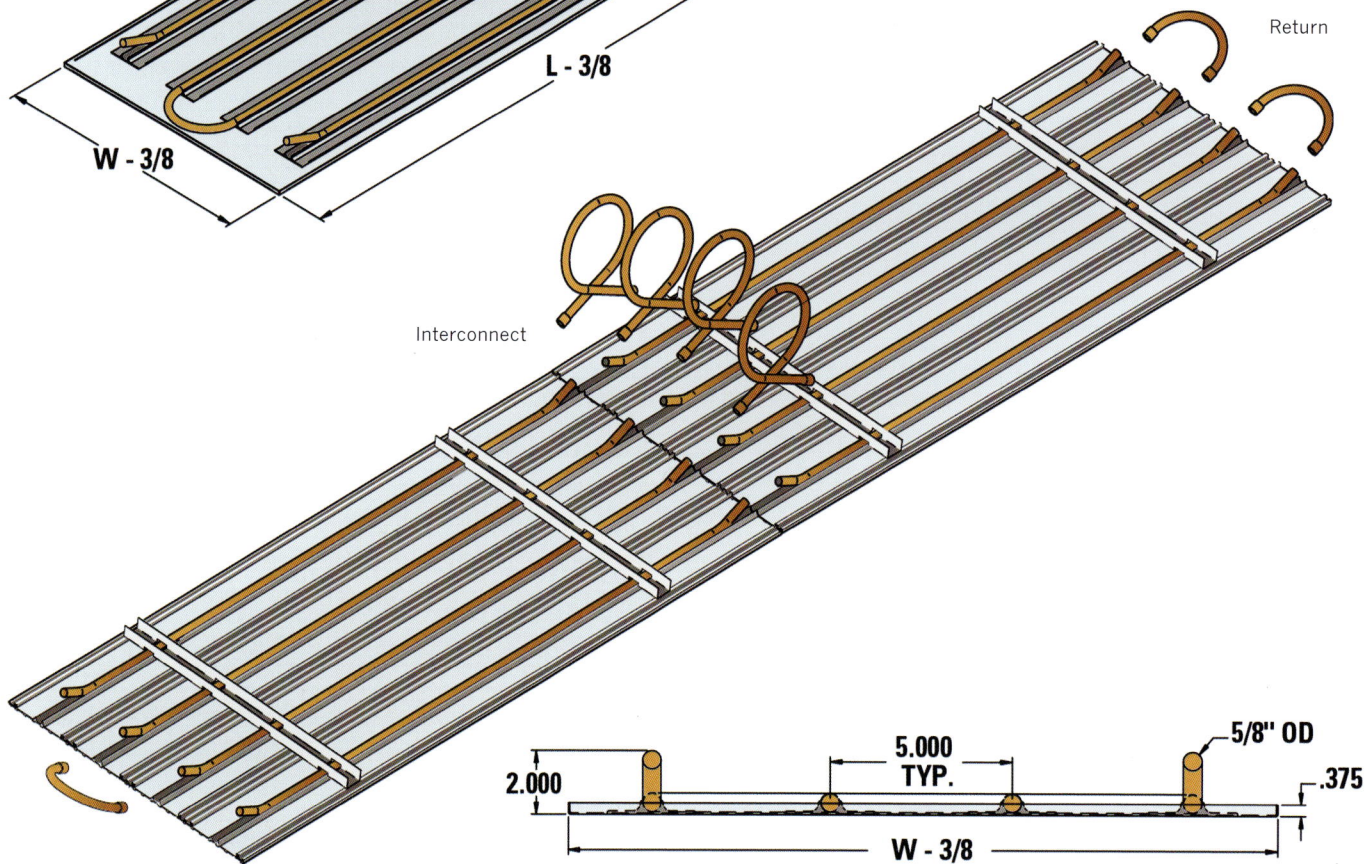

L - 3/8

W - 3/8

Return

Interconnect

2.000

5.000
TYP.

5/8" OD

.375

W - 3/8

GLUMAC'S FIRST RADIANT FLOOR:
Eugene, Oregon's Wayne L. Morse Federal
Courthouse project represented an ideal
candidate for radiant heating and cooling
in 2001. In-slab systems incorporate
HDPE tubing cast in concrete, using
water as the transport medium to heat or
cool the floor surface above. To alleviate
concerns about condensation on the
floor, mechanical engineers included
custom control sequences with sensors
embedded in the floor.

DESIGN PRINCIPLE 4

MAXIMIZE **DISPLACEMENT** **VENTILATION** OPPORTUNITIES

Low velocity/low temperature air supplied at or near floor level – utilizing natural convection forces to push warm air generated by occupants and equipment up and out of a building volume for efficient space conditioning.

Glumac continues to apply established MEP technologies like displacement ventilation in new and creative ways. In a commercial or institutional setting, these systems supply a "cool pool" of air in a space at floor level; through natural buoyancy, the air warms up and rises gently, attracted naturally to any heat sources – including people, computers and lighting – while maintaining comfortable conditions for occupants. Typical system components include the air handling unit, environmental air space of 12 to 16 inches (0.3 to 0.4m), and either wall-, corner- or floor-mounted diffusers. In open office environments, these systems utilize raised access flooring to distribute air and facilitate cabling and power changes for areas with high turnover rates. Displacement ventilation also performs well in semi-conditioned spaces that feature high ceilings, such as building lobbies and atriums.

Auditoriums represent another successful application for Glumac designers. Here, theatrical lighting creates the biggest space load, requiring that conventional overhead air systems produce additional cooling to counter the effects of heat gain from multiple light fixtures in addition to the heat generated by the audience during a performance. Though designed to minimize acoustic levels, most traditional diffusers also do not create adequate air flow, so temperatures vary widely from seat to seat within an auditorium. Instead, supplying air at the floor level – between 63°F and 65°F/17.2°C and 18.3°C – achieves stratified temperatures in a space. For example: occupied zones, at six feet and below, remain at desired temperatures, approximately 74°F to 76°F/23.3°C to 24.4°C; warm air migrates upward with temperatures reaching 85°F to 90°F/29.4°C to 32.2°C at the ceiling, where it is then extracted along with the heat from theatrical lighting, etcetera.

UNDERFLOOR AIR FOR THE OFFICE ENVIRONMENT

The use of raised access flooring to distribute underfloor air provides an affordable solution for reconfiguring office layouts in open floor plans. In contrast to conventional

QUICK FIX: Glumac's mechanical design for the City Development Center in 1999 resulted in the first large-scale raised access floor installation in Portland. Just two weeks prior to occupancy, the city chose to commingle departments, calling for reconfiguration of approximately 150 workstations. Hoffman Construction, the project's general contractor, estimated the added cost to relocate these workstations at just several thousand dollars – a savings to the City of more than $100,000 compared to a conventional building layout.

BEFORE: VESTAS HEADQUARTERS DISPLACEMENT VENTILATION

1 Outside air intake
2 Air handlers
3 Raised access floor
4 Underfloor air distribution
5 Floor diffusers

6 Atrium with stack ventilation
7 Skylight building relief with fan assist
8 HVAC return wall grill with sound trap

SIMPLE PHYSICS: Underfloor air distribution figures prominently in the LEED Platinum renovation of a Portland landmark for Vestas, a leading manufacturer of high-tech wind power systems. In the original HVAC layout, designers discovered return air inlets were placed too close to the perimeter, drawing heat from the skylight and creating excessive heat for upper floors. Glumac's computational fluid dynamic (CFD) models confirmed this.

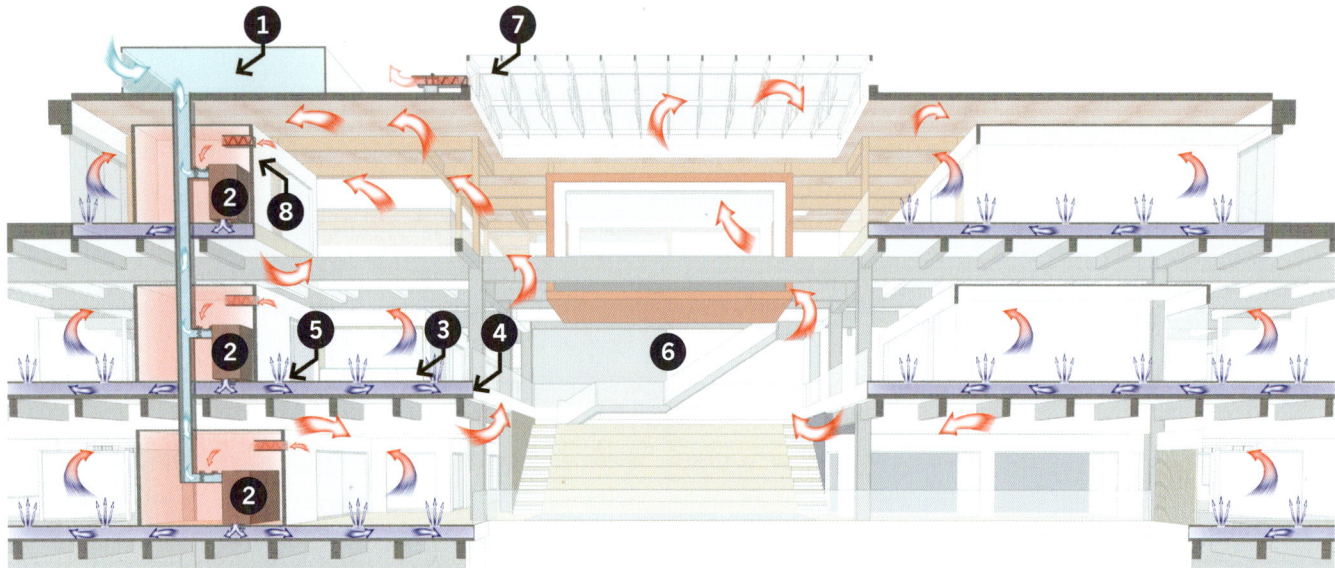

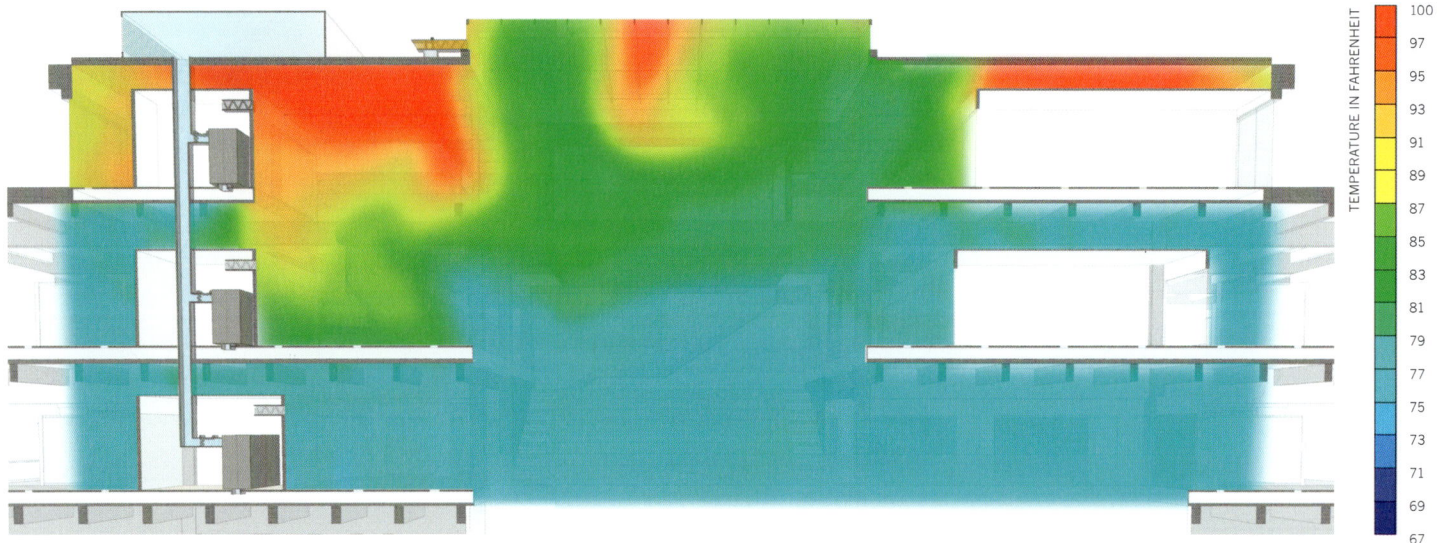

TEMPERATURE IN FAHRENHEIT

100 97 95 93 91 89 87 85 83 81 79 77 75 73 71 69 67

AFTER: VESTAS HEADQUARTERS DISPLACEMENT VENTILATION

1 Outside air intake
2 Air handlers
3 Raised access floor
4 Underfloor air distribution
5 Floor diffusers

6 Atrium with stack ventilation
7 Skylight building relief with fan assist
8 HVAC return wall grill with sound trap

CFD SOLUTION: In Glumac's revised HVAC schematic, designers moved return air inlets close to the skylight to solve the heat buildup problem. Again, CFD modeling confirmed these changes would improve air distribution at target temperatures throughout interior spaces.

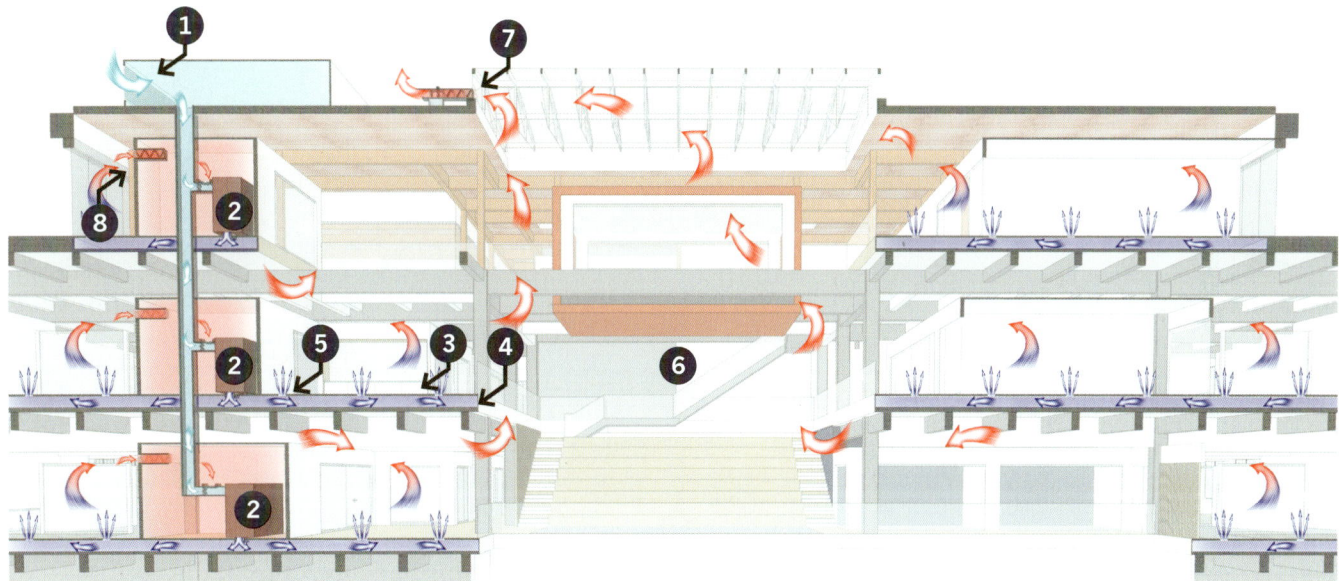

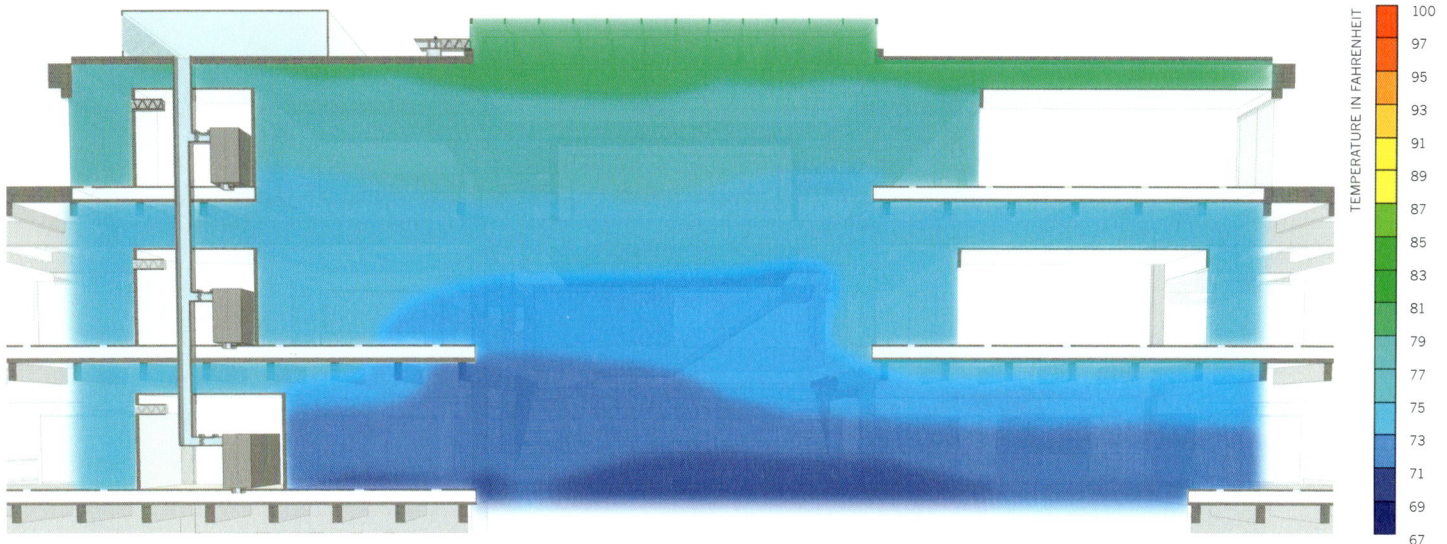

TEMPERATURE IN FAHRENHEIT

100
97
95
93
91
89
87
85
83
81
79
77
75
73
71
69
67

VESTAS HEADQUARTERS NIGHT COOLING

1 Outside air intake
2 Air handlers
3 Raised access floor
4 Underfloor air distribution
5 Floor diffusers

6 Atrium with stack ventilation
7 Skylight building relief with fan assist
8 HVAC return wall grill

AIR CHANGE: Opening in 2012, the 172,000 square foot (15,979 square meter) office will utilize its extensive thermal mass and a night purge system as the predominant means of cooling within the central atrium and open work spaces across five floors.

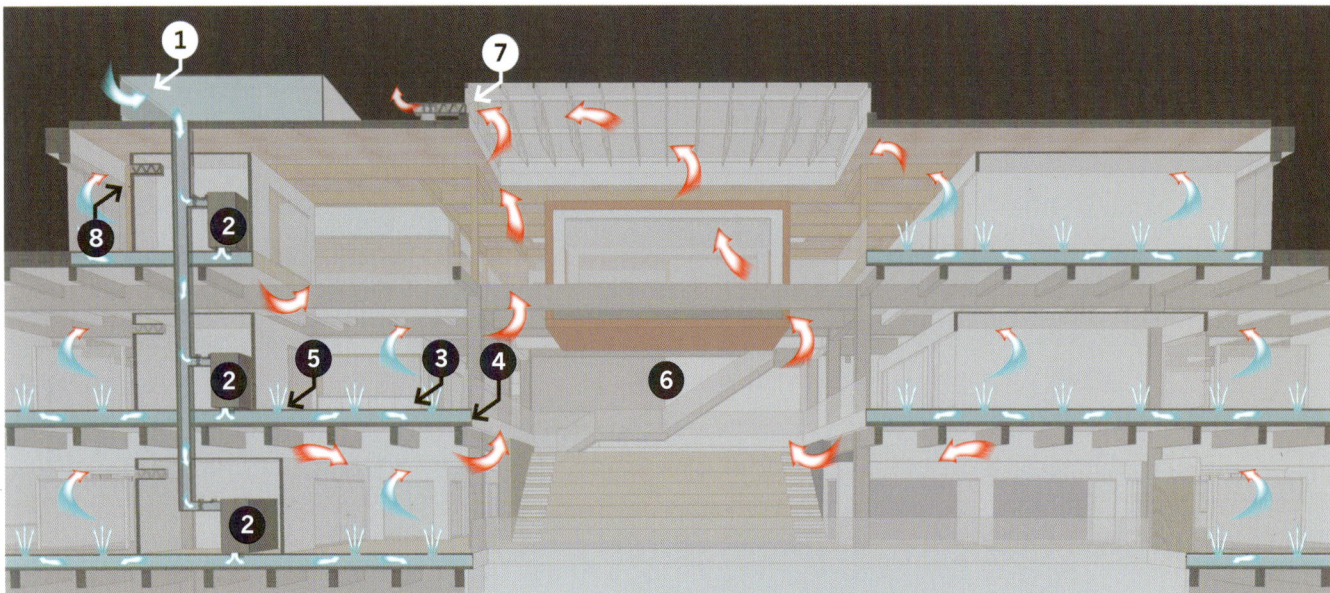

ductwork schemes, electrical outlets and air outlets are relocated simply by unscrewing and moving floor tiles. Additional benefits of these systems include reduced energy consumption and greater flexibility for individual temperature control. While some occupants prefer a cooler personal space, others want it warmer; each floor diffuser therefore contains an adjustable air flow control, so that every occupant may increase or reduce the amount of cooling air in their space.

On projects without raised floors, Glumac promotes the use of vertical displacement diffusers – placed in corners, free-standing, or mounted into walls and tied into the building's duct system. In each case, the displacement ventilation system delivers low velocity air across the space (e.g., lobbies and corridors), picking up heat from people and equipment and then rising to the ceiling.

AIR SIDE ADVANTAGES

Due to higher supply temperatures, displacement ventilation systems also take fuller advantage of a building's economizer cycle (the ability to cool with outside air), operating many more hours a year at 60°F to 65°F/15.6°C to 18.3°C – versus a conventional system that delivers air at 52°F to 55°F/11.1°C to 12.8°C. These devices, integrated into a central air handling system, bring in larger quantities of outside air to heat or cool

spaces and equipment while exhausting the return air. This "free cooling" approach can lead to significant energy savings. In San Francisco's climate, as an example, 4,165 hours of the year fall below 55°F/12.8°C (with 980 hours between 7:00 a.m. and 7:00 p.m. Monday through Friday). In addition, 7,671 hours of the year fall below 65°F/18.3°C (with 3,952 hours between 7:00 a.m. and 7:00 p.m. Monday through Friday).

PROCESS/TOOLS

Glumac reviews several criteria before recommending a displacement ventilation scheme. For example, does the project feature an open office, with an expected high churn rate? How will space be used: does the building include large conference rooms, with program areas open from 9 a.m. to 5 p.m. or do they operate late – even 24/7? Designers also weigh the following points:

» *Higher discharge air temperatures*: This approach calls for a completely different type of design thinking. Where conventional systems typically introduce air at a cooler 55°F/12.8°C to condition entire spaces, discharge air via displacement diffusers enters at a minimum of 63°F/17.2°C to condition only occupied spaces (up to six feet/1.8m); otherwise, lower temperatures may tend to draft down to ankle level and create an uncomfortable environment.

» *Location (proximity to outdoor air temperatures and humidity)*: To dehumidify a space requires cooling the air to 55°F/12.8°C, so this technology is not well suited for humid climates without a return air bypass or a dedicated outside air system.

» *Decoupled systems*: Displacement technology works well in tandem with chilled beams/overhead radiant at the ceiling level. With this combination, displacement ventilation supports the base (or constant) loads of the space, while the overhead systems provide supplemental heating or cooling.

» *Environmental air space*: For raised access floors, the ideal design space allowed for air movement varies from 12 to 16 inches/0.3 to 0.4m, depending on the floor plate. While there are exceptions, any less vertical space may require additional air handling units to move air effectively beneath the floor; likewise, anything higher than 16 inches/0.4m may require seismic restraints according to code.

FURTHER DESIGN FACTORS

Beyond programming and structural considerations, cost becomes a critical factor in determining the feasibility of these systems. Raised floors, for example, cost $5 to $10 more per square foot to install (estimates vary according to vendor, type of flooring, grilles, etcetera.) than conventional overhead systems. However, the benefits may far outweigh the added expense in terms of improved ventilation efficiency and savings over time with high office churn rates.

In addition, underfloor systems studies consistently demonstrate improved personal comfort and better indoor air quality. While overhead ductwork promotes mixing of air throughout an office, distributing germs, dust and dirt, the "single pass" design of displacement ventilation systems – which condition only the occupied space as air rises up – can also mean higher productivity and fewer sick days.

DISPLACEMENT VENTILATION

1 Raised access floor
2 Adjustable floor diffusers
3 Supply air duct
4 Underfloor air distribution plenum
5 Return air ceiling grille
6 Return air plenum in suspended ceiling
7 Return air duct

COOL AIR RISING: A displacement ventilation system, supplies conditioned cool air from an air handler unit through a low induction (in this case, floor-mounted) diffuser. Cool air spreads through the floor of the space and then rises as the air warms – due to heat exchange with heat sources in the space (e.g., occupants, computers, lights). The warmer air has a lower density than the cool air, creating upward convective flows known as thermal plumes. Warm air exits the zone at the ceiling height of the room.

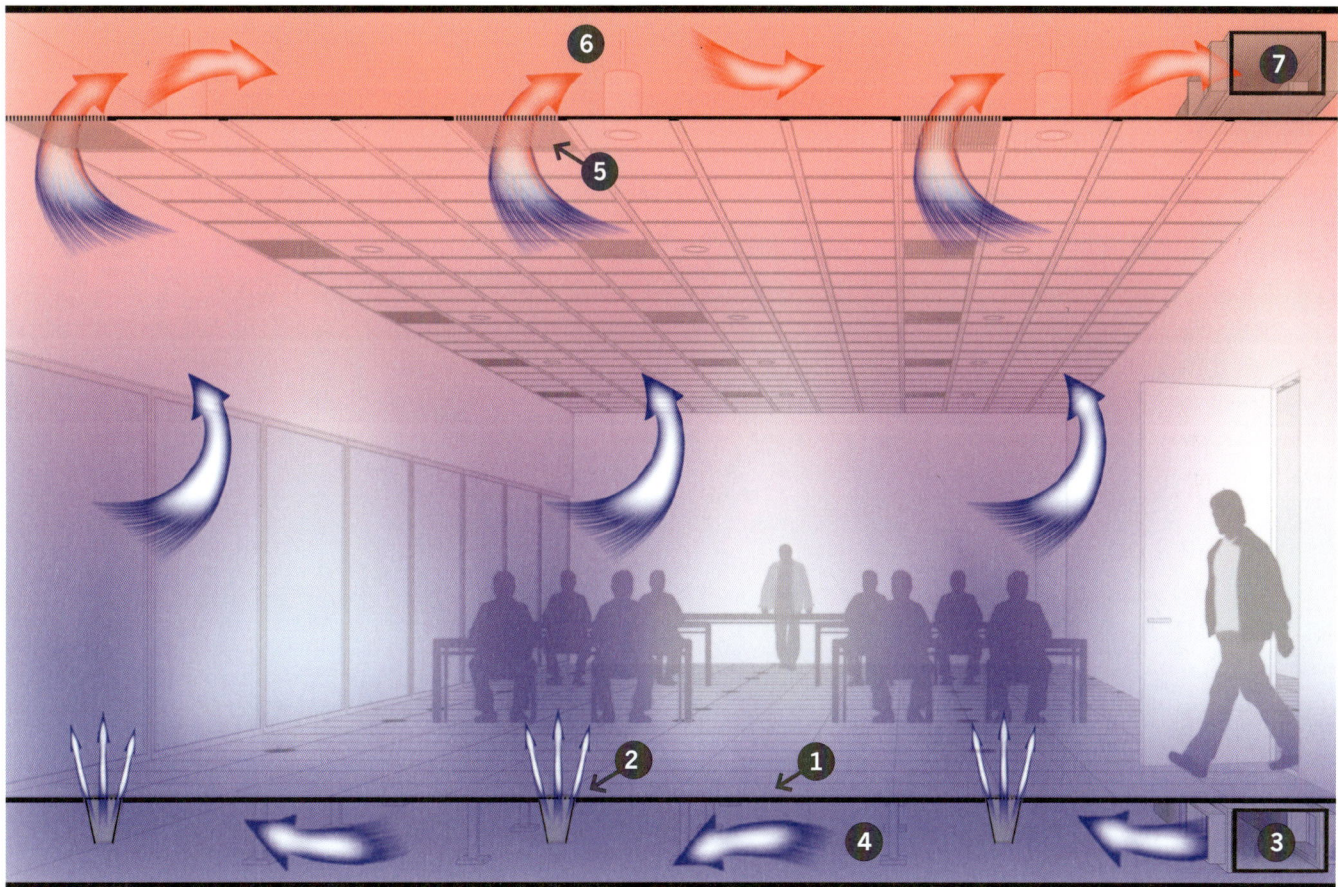

APPLY **NATURAL** **VENTILATION** SOLUTIONS EARLY AND OFTEN

The movement of fresh air – using the natural forces of wind, temperature differences and buoyancy – within an indoor space through openings in the wall of a building, without the use of a fan or other mechanical system.

DESIGN PRINCIPLE 5

HIGH PERFORMANCE: Day or night, the new Health and Wellness Center at Western Oregon University maximizes natural ventilation to continually refresh air through a series of design strategies.

Fresh air serves as one of the most underutilized and yet one of the simplest measures for conditioning facilities, if ambient air quality and temperatures are conducive. The many advantages of natural ventilation include flexibility (open or close windows), environmental control at the occupant level, and passive energy savings (versus forcing air through ducted systems). Common design schemes promote either wind-induced *cross ventilation* – which relies on orientation, narrow floor plates and large openings – or *stack ventilation* – through natural buoyancy based on temperature differentials from low intakes and high exhaust.

As with daylighting, Glumac's designers continually strive to introduce natural ventilation into buildings. Their mindset reflects a *hybrid* approach: sizing the HVAC system optimally for air conditioning demand while incorporating passive features into every design to take advantage of mild seasonal and climatic opportunities. They often compare this approach to operating a sailboat in the wind: determining whether to turn on the motor or just trim the sails. Gaining owner buy-in becomes critical to success with natural ventilation, as well as addressing several potential roadblocks to this strategy. These considerations include:

»   *Conducive climate*: Do summer high and/or winter low temperatures make this a practical solution?
»   *Building loads*: Loads in server rooms, computer labs and similar spaces may be too high to ventilate naturally
»   *Building geometry*: Projects with narrower footprints allow effective cross-ventilation; those with atriums and other architectural "chimneys" maximize the potential for stack ventilation
»   *Cost*: Operable windows generally cost more than fixed windows

NATURAL VENTILATION IN THE STATE OF OREGON NORTH MALL OFFICE BUILDING

1 Natural ventilation inlet integrated into perimeter bench
2 Natural ventilation relief integrated into south facing skylight and clear-story windows
3 Relief Air from office space air-handling unit (AHU) recirculated in atrium through floor diffusers
4 Office space AHU with relief air ducted to atrium

TRANSITIONAL SPACE: The North Mall atrium recirculates relief air (78°F/25.6°C) from air handlers serving office space throughout the 115,000 square foot, three-story building. Strong airflow and slightly higher temperatures within the atrium yield comfortable conditions year round. The recirculation system, combined with the natural ventilation scheme, also produces energy savings by optimizing the conditioned air and relying on the natural buoyancy of warm air rather than mechanical equipment to ventilate the space.

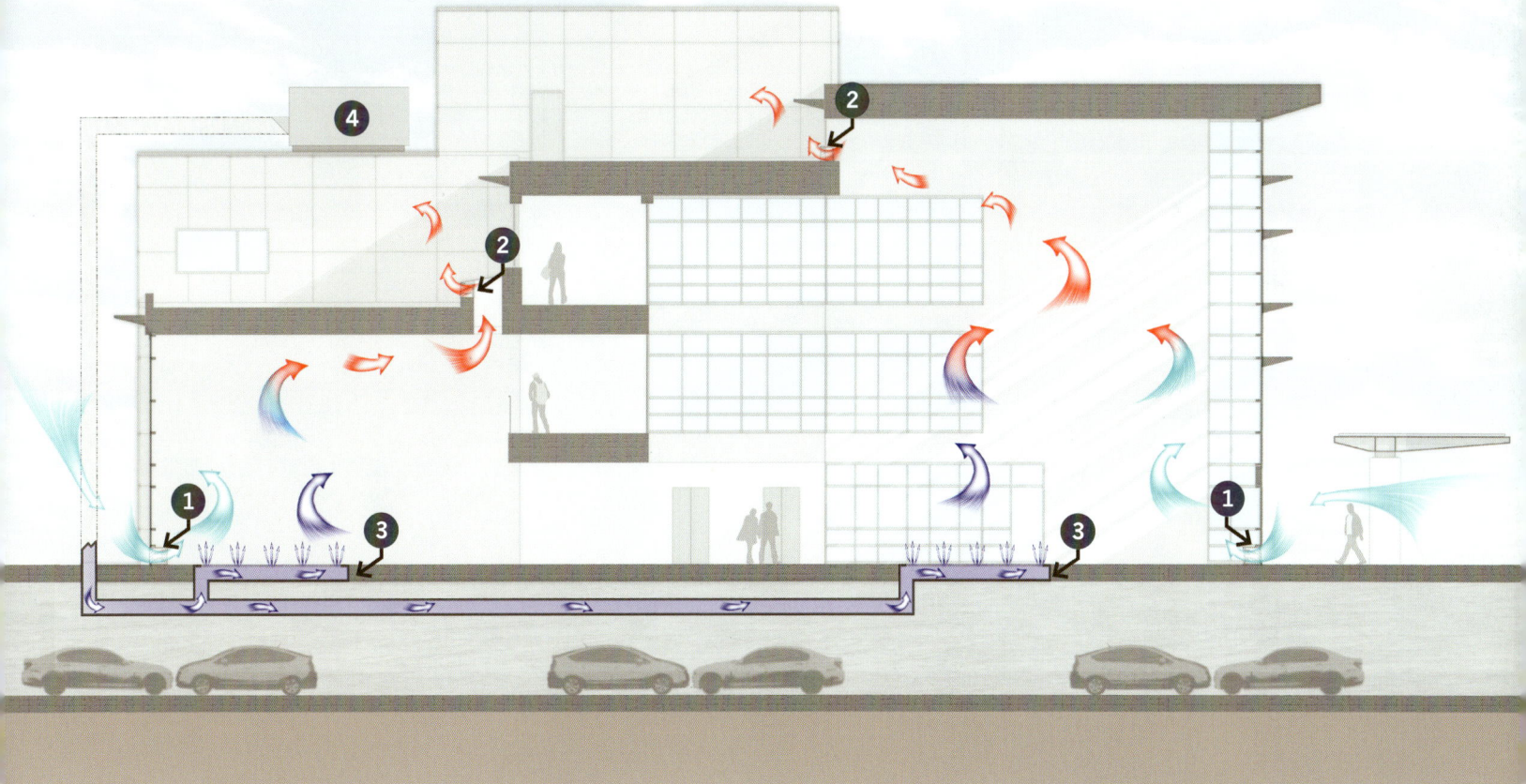

EXIT

NORTH MALL: Designed as a Sustainability Pilot Project in 2004, the State of Oregon North Mall Office Building (in Salem) utilizes its three-story atrium for natural ventilation without air conditioning — while serving as a connection and buffer zone between three buildings in the complex. Glumac provided mechanical and electrical, technology integration and sustainable design services.

BY PASSIVE MEANS FIRST

Glumac believes natural ventilation as a strategy should be at the forefront of possibilities for any sustainable design project. Resolving concerns up front, such as outdoor air quality, noise, security or allergens, may help the owner more fully embrace this concept to create a better building. Early discussions with the architect and other members of the project team can also streamline the design process. For example, optimizing the building envelope reduces internal loads to allow passive conditioning at higher temperatures. Shaping the building for airflow means narrower floor plates for cross ventilation and including stack elements, such as atriums, to promote fresh air strategies such as night purging. Also, adding simpler building controls may improve the operation of some passive ventilation schemes.

Operable windows represent perhaps the most visible symbol of a naturally-ventilated building, and are ideal for perimeter spaces. By providing occupant control, manual windows work well for individual offices, conference rooms, residential towers and hotels. Yet they can be problematic for open offices and in public spaces like lobbies and galleries, as comfort levels differ from person to person. Manual operable windows result in significantly higher energy consumption if used improperly. Glumac points to the example of one academic building, engineered for natural ventilation to reduce cooling energy:

occupants sometimes leave windows open after work hour, causing the HVAC system to compensate to keep the space warm · this causes heating costs to spike up to five times higher than other buildings on the same campus.

TECHNOLOGY AS APPROPRIATE

Operating guidelines and other procedures for occupants become more essential with passive designs, as do automated control systems. Several Glumac projects feature micro-switches to shut off the HVAC system automatically when windows open and during off-hours. Use of motorized windows, louvers and dampers also facilitate overnight purging as part of a larger control scheme to freshen indoor air quality and cool a building naturally following the previous day's heat-producing activities and increased solar loads in summer. Typically, controls open windows and other intake points automatically in the evening or early morning hours between 2 a.m. and 4 a.m., exhausting stagnant air through high openings. Again, early design decisions shape building loads, programming and overall architecture to make night purges as effective as possible.

PROCESS/TOOLS

Glumac engineers base their natural ventilation designs on intuition and experience, all coordinated closely with the architect. Initial designs are then tested using computer-based CFD (computational fluid dynamics) modeling to predict air flow

YEAR-ROUND COMFORT: Glumac designers incorporated natural ventilation into benches along both sides of the building's central atrium: providing both a heat sink and source of cooling according to season – yet another innovative detail earning the project its LEED Gold rating.

within spaces more precisely and help guide overall decision making.

Designers also factor in several general guidelines and pre-conditions:

» *Space ventilation*: The design must ventilate a minimum of 4 percent of floor area to meet code
» *Cross ventilation* (blowing air from one side to another): Limit exterior-to-exterior distance to less than or equal to five times the height of the space; for example, if the space is 10 feet (3.048 m) high, limit the exterior-to-exterior distance to 50 feet (15.24 m)
» *Pressure drop*: Limit motorized openings to very small pressure differentials of less than 0.05 inches (0.127 cm) to induce stack ventilation
» *Window openings*: Design large openings and place intakes on the windward side of a building, with outtakes/reliefs on the leeward or discharge side
» *Building loads*: Calculate to keep equipment, lights, etcetera at no more than 1 W/sf (10.76 W/sm) so that internal loads do not mandate mechanical cooling systems

FURTHER DESIGN FACTORS

Natural ventilation depends in part on the cooperation of building occupants for its success. Passive design elements, by nature, require broader tolerances – particularly with indoor air temperatures varying from 68°F to 78°F (20°C to 25.56°C) to achieve ideal comfort levels in most climates. As a result, occupants should dress appropriately – warmer in winter and cooler in summer. Glumac also continues to experiment with behavioral tools like the Center for the Built Environment's "red light/green light" system[1], used to notify occupants to open or close windows. These lights may be located prominently at the perimeter of a space or programmed as a small dot in the corner of employees' computer screens: red means temperatures are too hot or cold to naturally ventilate, while green indicates optimal conditions.

1. "Behavior and Buildings: Leveraging occupants to improve energy and comfort," *centerline*, Newsletter of the Center for the Built Environment at the University of California, Berkeley, Summer 2010.

ARCHITECTURAL INTEGRATION OF PERIMETER BENCH DETAIL

NATURAL VENTILATION IN THE STATE OF OREGON NORTH MALL OFFICE BUILDING

In summer, louvers inside the bench open with the heaters off, allowing a maximum amount of fresh outside air to naturally ventilate the atrium.

In winter, louvers inside the bench close with the heaters on – this design warms a minimum amount of cold fresh outside air, while heating and recalculating the cold air at the perimeter windows of the atrium.

ADD **GEOEXCHANGE** AS AN ALTERNATIVE ENERGY SOURCE

Utilizing the constant, year-round temperature of the earth (or groundwater, lake or pond water) as both a heat source (in winter) and heat sink (in summer) via an electrically-powered heat pump for space conditioning.

GEOEXCHANGE HEATING/COOLING SYSTEM

1 Bore field with geothermal wells
2 Supply manifold
3 Return manifold
4 Ground loop pump
5 Heat exchanger
6 Water to air heat pump
7 Water to water heat pump
8 HVAC system pump
9 Bypass line

10 Ventilation air outlet
11 Thermostatic control valve
12 Thermally activated concrete with
     embedded radiant pipes

GROUND TO AIR: Offering highly efficient energy transfer, a vertical closed loop system extracts heat from air in the building (in summer), transferring it through circulating fluid via HDPE piping and back into the earth. In winter, the reverse happens, as geoexchange technology makes it possible to absorb heat from the earth, moving it inside for warmth.

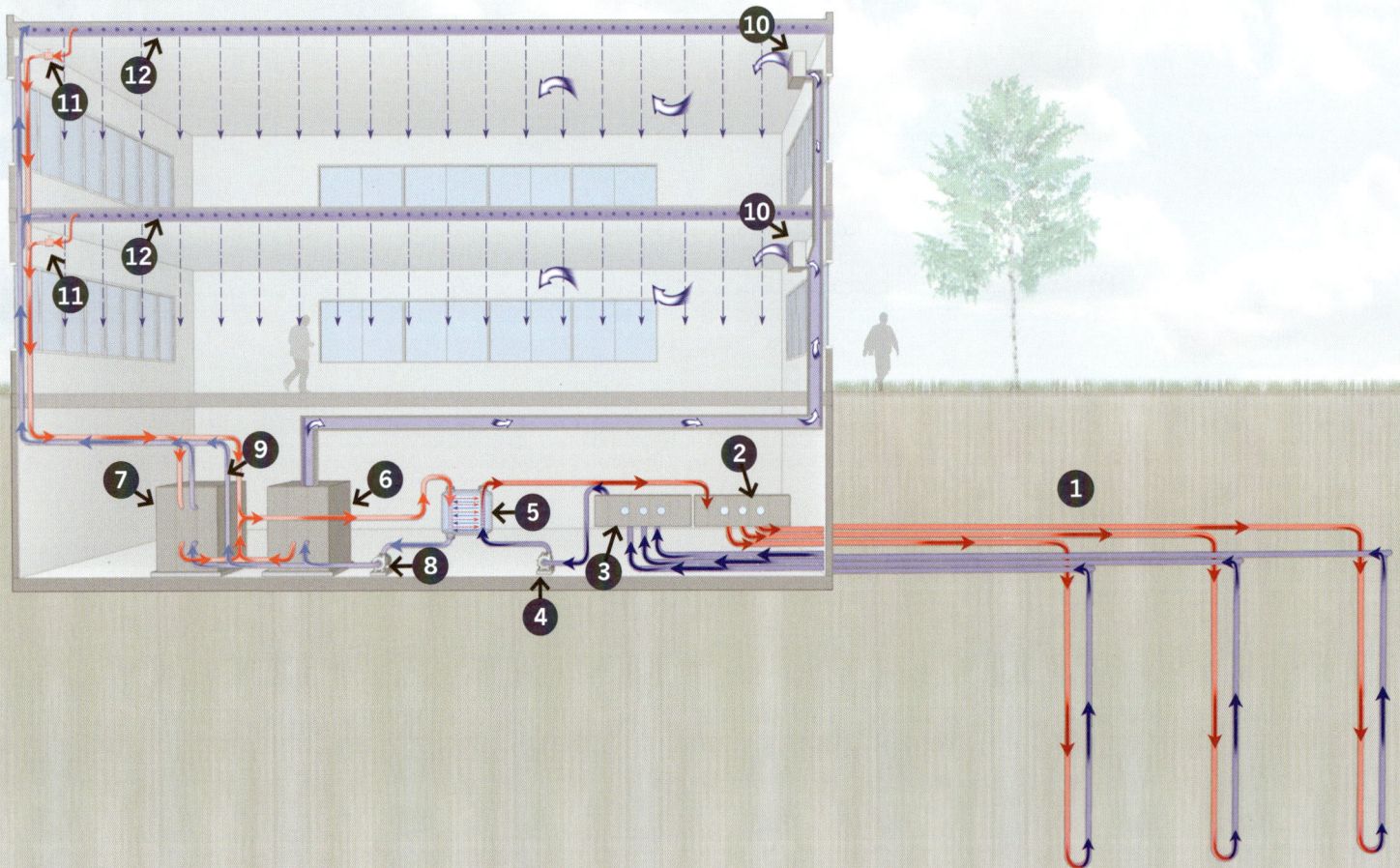

The earth provides the perfect heat exchange medium: harnessing constant, year-round temperatures to deliver highly efficient heating and cooling energy for all types of building projects. Although commonly referred to as geothermal, earth-coupled heat pump or ground-source heat pump technology, the industry now prefers the term "geoexchange" systems for this alternative source of energy.

As designed and installed, geoexchange carries a higher initial cost than standard HVAC – a cost that varies according to project size, scope and drilling conditions for each locale. Yet these systems also operate at much higher year-round efficiencies than conventional systems, typically requiring 40 to 70 percent less energy and lower maintenance costs while resulting in longer equipment life and a smaller carbon footprint. The investment on most installations also realizes a payback between five and eight years. Rising fossil fuel costs, LEED® points for building efficiency, and new government incentives and utility rebates also contribute to the lowest life-cycle cost of any heating and cooling technology. For more than a decade Glumac has recommended geoexchange systems for schools, retail, office buildings, single- and multi-family residential and light commercial buildings. Ideal applications include spaces that do not operate 24/7 (like fire or police stations), so building loads allow temperatures within the earth/ground loop piping to settle out and recover during off-hours.

SYSTEM CONFIGURATION

The optimal geoexchange system design for each project is dictated by building loads, acreage available, geotechnical (soil/rock) conditions and other factors. Earth exchange configurations fall into four main classifications: open or closed loop, vertical or horizontal layouts. An open system involves groundwater (vertical wells) or lake/pond water as the supply source for heating and cooling. This approach can be very cost-effective, assuming high water quality, sufficient water quantity, discharge location available, and compliance with all local codes and standards.

Glumac generally prefers closed loop systems due to cost, permitting requirements and environmental concerns. In these systems, fluid circulating through buried plastic piping (a series of ground loops) absorbs heat from the earth in winter, delivering it indoors via a heat pump for warmth; in summer, the reverse occurs, as the system extracts heat from air in the building, transferring it through the fluid and piping and back into the earth. Vertical closed loops typically consist of four-inch (10 cm) diameter bores, drilled between 200 and 350 feet (61 and 106 m) depending on application and climate, and spaced 15 to 20 feet (4.6 to 6.1 m) off center (o.c.). Installers then place one-inch (2.5 cm) flexible HDPE pipe into the bore holes, filling and sealing the annular space with thermally-conductive grout. Water serves as the preferred heat exchange fluid, moving continuously through a vertical system's circuits (either "individual loops" or "common loop") and runouts. In colder climates with larger heating loads, the water contains a small percentage of glycol to prevent freezing.

Horizontal layouts – configured either as a closed loop or an open system that utilizes lake or pond water – represent another option. This design requires more land and, therefore, approximately twice as much piping. Loops/supply lines must be buried five or six feet (1.5 or 1.8 m) apart and a minimum of five feet (1.5 m) deep, as a buffer against ambient temperatures on the ground's surface.

THE BUILDING INTERFACE

Another advantage of geoexchange: the water loop (or "earth exchanger") also serves as the system's condenser – with no equipment actually placed *outside* the building. This technique is preferable for landscaping, is safer and quieter, and requires no maintenance. Glumac's recent geoexchange design for the Sacramento Housing & Redevelopment Agency relies on a closed vertical loop system to heat and a cool low-income housing development. A water-source heat pump, located within the storage closet of each unit, looks and functions much like a standard air-to-air heat pump, with forced air distribution/ductwork resembling a conventional HVAC system. In addition, these water-to-air heat pumps feature an extended range, enabling equipment to operate at lower temperatures (i.e., 50°F/10°C). An open system, which Glumac specified for a golf course clubhouse, pumps water from an adjacent lake and includes a filtration system and settling tank. In turn, that water is returned to the lake. A well system would typically utilize injection wells for reintroducing ground water to the aquifer.

PROCESS/TOOLS

Glumac's role on a geoexchange project begins with engineering the system layout to influence the design and construction of the bore/well field. Providing load calculations and energy modeling data early enables the driller or hydrogeologist to more accurately determine the system length and configuration to achieve optimal

TAPPING THE EARTH: A typical geoexchange loop provides one ton (3.5 kW) of cooling – utilizing bores drilled under a building, or more typically, in a landscaped or parking area. To support a 100,000 square foot (9,300 square meters) office building requiring 200 tons (700 kW) of cooling, the system would feature 200 bore holes – a field approximately 200 feet by 200 feet (61 meters by 61 meters).

heat transfer. Larger systems also require test bores to gather thermal conductivity data based on soil and rock strata. A few additional facts to consider:

» A typical geoexchange well/vertical loop provides one ton (3.5 kW) of cooling

» The linear feet of piping required per ton to support a system can vary greatly – and depends on each project's specific load and site characteristics

» Generally, a horizontal layout/loop must be *twice* the length of a vertical system to achieve equivalent heating/cooling capacity

FURTHER DESIGN FACTORS

While the interior (heat pump, ductwork) elements of a geoexchange system generally match the installed costs of conventional HVAC, its exterior elements (bore drilling, pipe, grout, pumps, etcetera) call for a bigger upfront investment. In addition, these hydrogeologic components require a potentially greater time commitment due to regulatory permits and the need for a timely, accurate assessment of the resources such as rock/soil conditions and ground water.

Selected projects may also utilize other traditional technologies to supplement geoexchange, particularly in areas with high boring costs or where heating and cooling loads are imbalanced. Cooling and heating profiles often require only 50 percent of peak loads for 80 percent of operating hours. This "hybrid loop" system can then optimize energy performance for most of the year, while shifting part of the load to a fossil fuel boiler or cooling tower for the warmest and coldest days.

FUTURE-PROOF ENERGY NEEDS WITH **PHOTOVOLTAIC SYSTEMS**

The direct conversion of sunlight into electricity – utilizing technology in which radiant (photon) energy creates a flow of electrons – to generate power by connecting individual solar cells in series and in parallel.

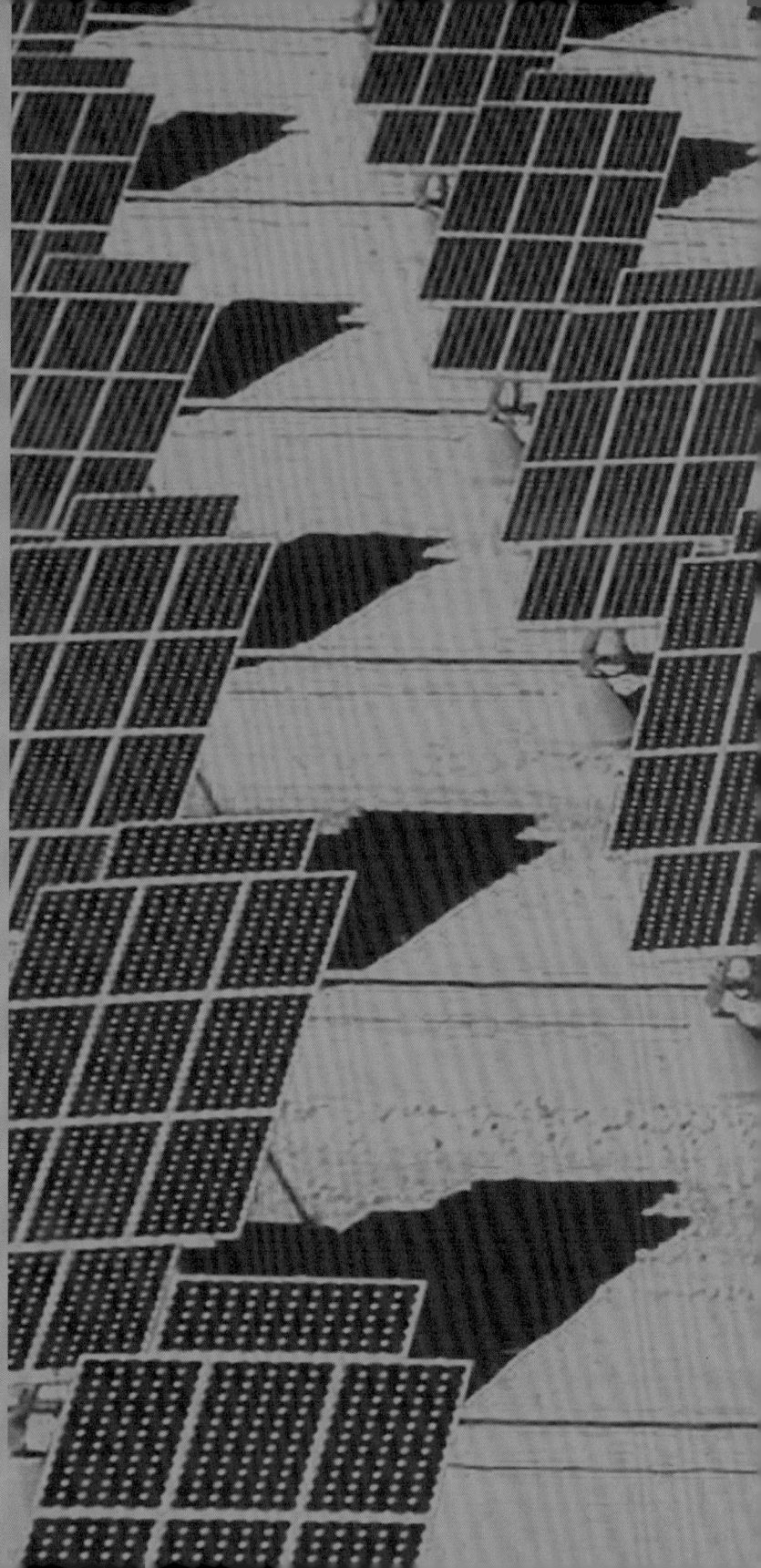

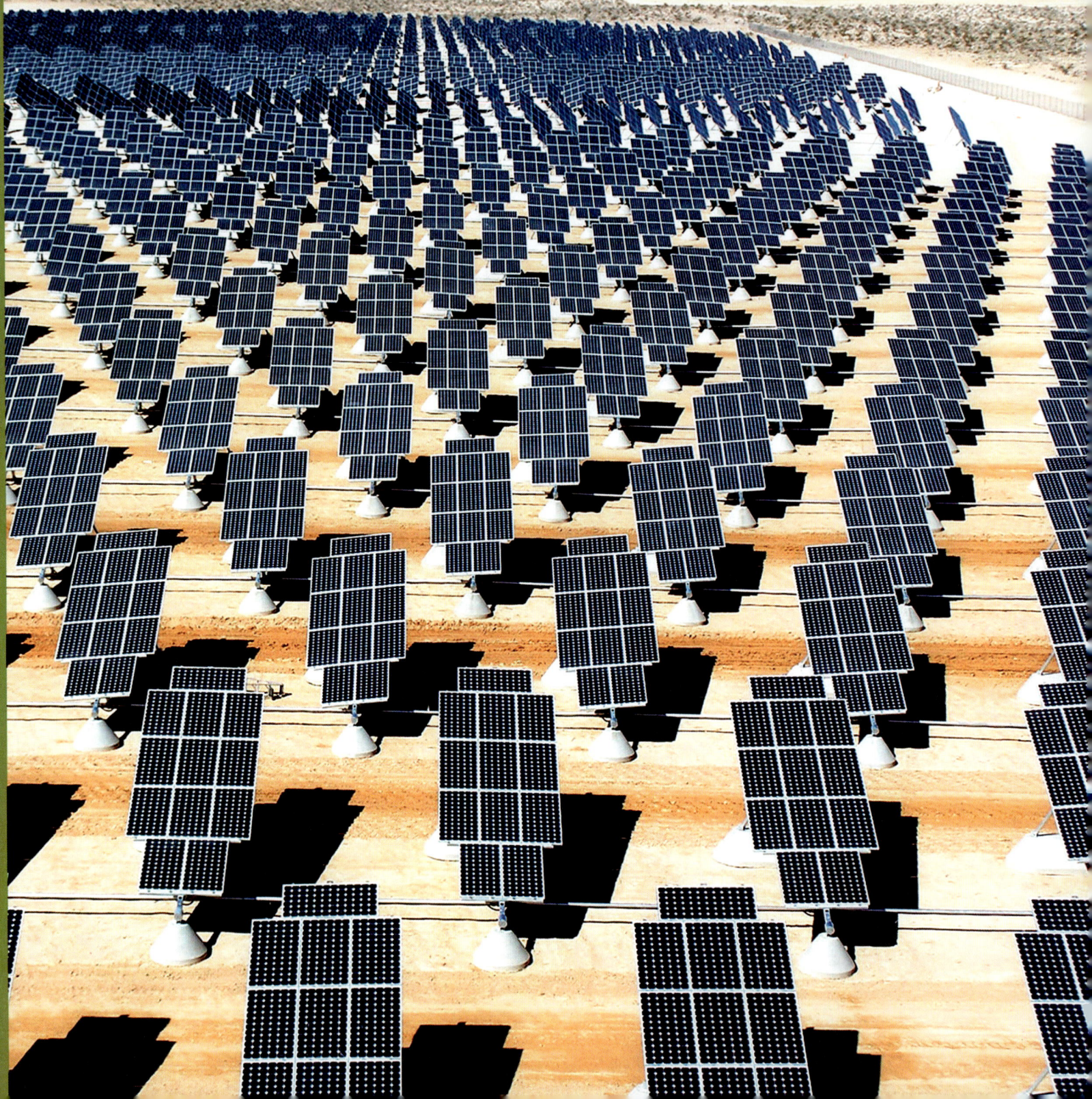

Solar electric – or photovoltaic (PV) – technology shows incredible promise in powering the world's buildings. The physics of PV have not changed: the energy of absorbed light transfers to electrons in the atoms of the PV cell's semiconductor material, converting sunlight directly into electricity. However, dramatic improvements in material chemistry allow those cells to produce energy from a much broader spectrum of light while making PV system junctions and other components more efficient. Today, the U.S. leads the world in R&D spending for solar technologies, with substantial investments in thin-film PV. The number of grid-connected PV installations as of Q1 2011 had grown 66 percent over Q1 2010[1]. The cost of producing PV, too, continues to decline as industry scales up manufacturing capabilities, while installation costs drop with the rise in more experienced, trained installers. In fact, the average cost of these systems decreased by more than 30 percent from 1998 to 2008 thanks to numerous national, state and local incentives[2].

Solar potential spans the continent and may be applied virtually anywhere. Glumac continues to specify and design building-integrated PV systems for a growing number of projects. While first cost (currently around $2.50/W installed for 2011) remains higher on average compared to conventional energy sources, tax credits and utility rebates can make these installations cost-effective.

SOLAR WORLD: Application of photovoltaic technology to projects for supplemental energy continues to grow – a market set to explode worldwide as the cost of PV panels declines quickly, making solar energy cost-competitive within a decade[3]. The price of each watt of peak capacity is expected to fall to about $1 within two years, compared to $2 in 2009.

1. U.S. Solar Market Insight™: Q1 2011, Solar Energy Industries Association (SEIA)® and GTM Research, 2011.
2. Wiser, Ryan and Galen Barbose, Carla Peterman, Naïm Darghouth. *Tracking the Sun II: The Installed Cost of Photovoltaics in the U.S. from 1998 – 2008*, Lawrence Berkeley National Laboratory, October 2009. http://eetd.lbl.gov/ea/emp/reports/lbnl-2674e.pdf.
3. "Ernst & Young UK solar PV industry outlook: The UK 50kW to 5MW solar PV market", June 2011

THE BREWERY BLOCKS: ROOFTOP PHOTOVOLTAIC SYSTEM

1 PV rooftop array
2 D.C. source circuit wiring
3 Branch circuit wiring
4 D.C. disconnect
5 Inverter
6 A.C. disconnect
7 Distribution panel
8 Bus plug-in circuit breaker
9 Electrical riser busway

PORTLAND PV: Even the Pacific Northwest offers viable solar potential to offset fossil-fuel based power. Glumac's photovoltaic scheme for the Pearl District's Brewery Blocks development combines rooftop and building-integrated photovoltaic systems, connected through a system of collectors and inverters to the electrical distribution system of one building – producing more than 20,500 kW annually.

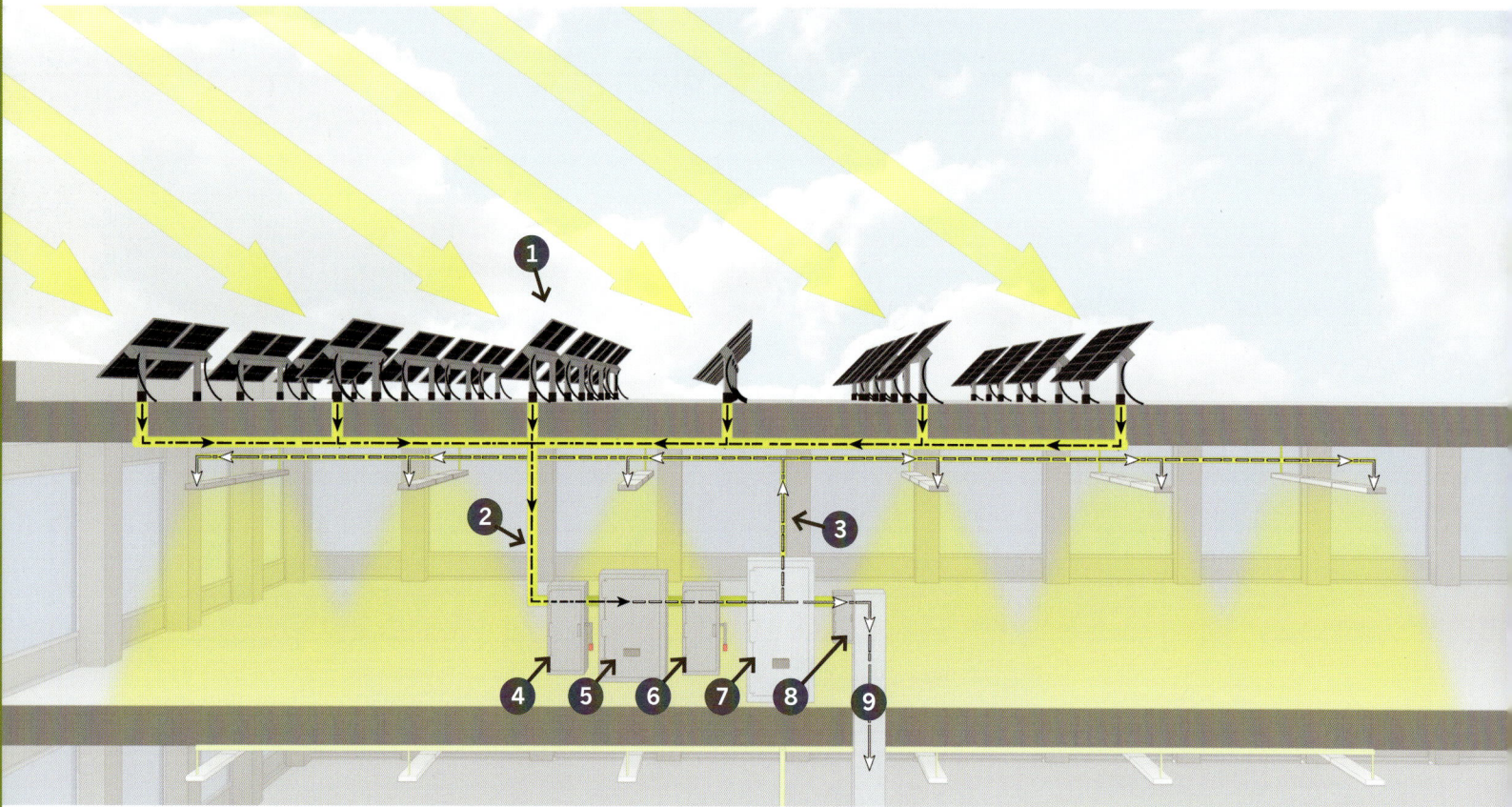

THE BREWERY BLOCKS: BUILDING-INTEGRATED AND ROOFTOP PHOTOVOLTAIC SYSTEMS

1 Rooftop PV array
2 Building integrated PV array
3 Area shown in detail illustration
4 Electrical riser busway
5 Main switchboard
6 Electrical service from utility grid

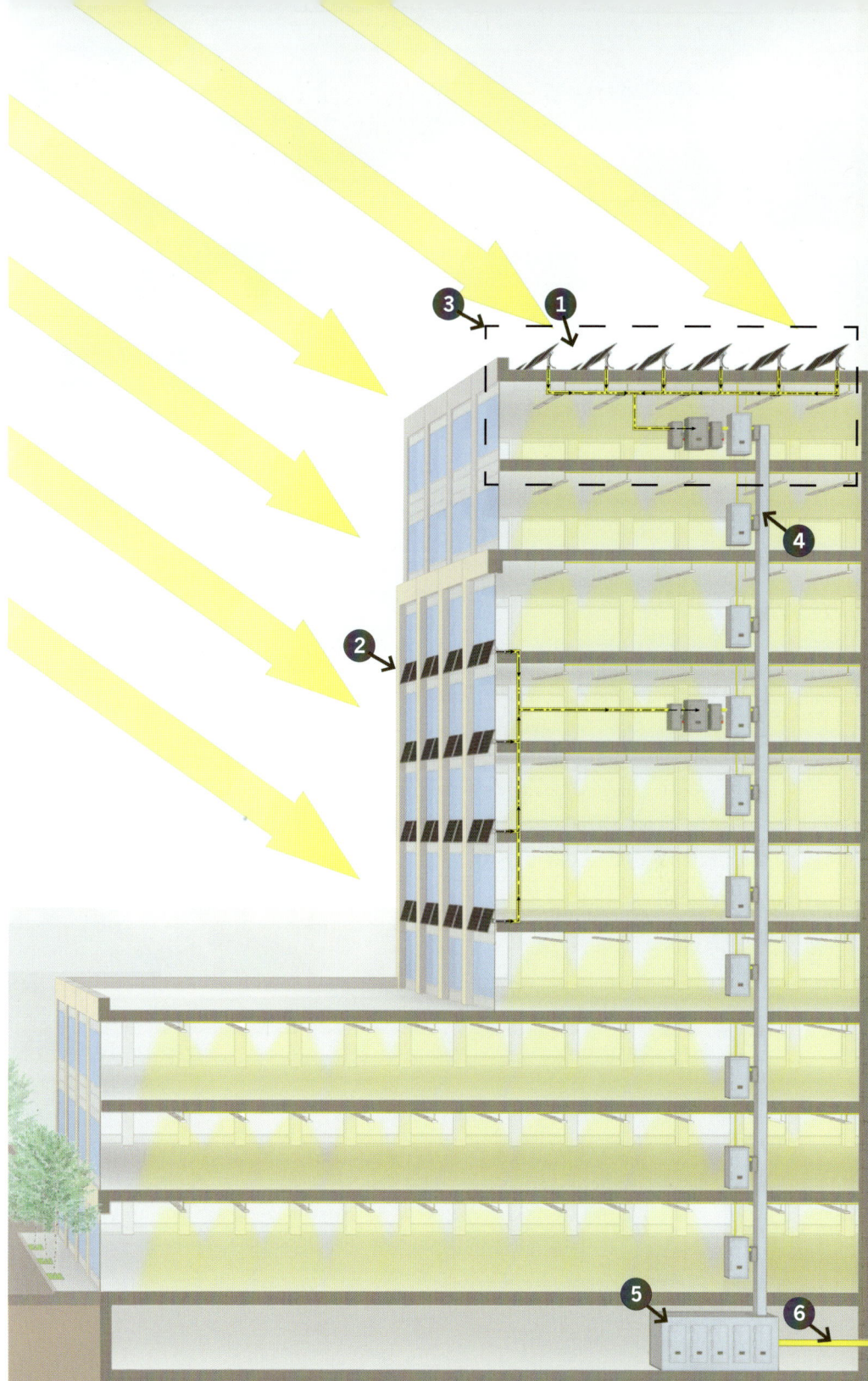

BUILDING INTEGRATED PHOTOVOLTAICS

While stand-alone PV systems represent a valid option, integrating photovoltaics into architecture can further optimize performance with several added advantages. Examples include:

» *Shading Canopies* provide a triple benefit of generating electricity, reducing cooling loads, and minimizing glare.

» Using photovoltaic panels for rooftop *Equipment Screening* provides a dual benefit of creating electricity and reducing the solar load on the roof, as well as screening equipment with a material that provides a visible symbol of sustainability.

» *Parking Garage Canopies* utilize a rooftop PV array to shade cars, reduce cooling load and create energy. Additionally, the panels can assist in rainwater capture for toilet flushing or irrigation; otherwise, the top deck of a parking garage, typically contaminated with oil and gas, presents a problem for water quality.

SOLAR CELLS, MODULES, ARRAYS AND THE BALANCE OF SYSTEM

The building blocks of a PV system, individual *solar cells,* vary according to crystallinity, bandgap, absorbtion and manufacturing complexity. Because a basic solar cell produces a relatively small amount of power (typically 1 or 2 watts), manufacturers connect cells together to form PV *modules* or *panels* that may extend up to 4 feet by 10 feet (1.2m by 3m) in size. In turn, installers combine and connect panels to form PV *arrays* of different dimensions and power output to meet varied, and growing, electrical requirements. As part of an array, panels may be fixed in place, generally facing south, or mounted on a tracking device to follow the sun.

Glumac designers give careful consideration, first, to connecting enough PV panels to achieve a terminal voltage of 380 to 400 volts. Next, they combine strings of 15 or 16 panels each to add capacity (in a typical scenario). While south-facing roofs and surfaces offer the most ideal orientation for energy production, it is also critical to mount and position PV panels so an entire array receives equal amounts of sunlight.

Balance of system (BOS) components consist of *mounting or tracking structures* for the PV arrays and *power conditioners*, which serve an essential function in processing the electricity produced by the PV system to meet energy load demands. For DC applications, regulators provide power conditioning; for AC loads, the equipment must include an inverter to convert DC electricity into AC power. In addition, *batteries* and a battery backup system store solar energy for use when the sun is not shining, during cloudy days or at night. Finally, *charge controllers* protect batteries from overcharging and excessive discharge. *Inverters* also serve a critical purpose in power generation, commonly applied in three ways: to support discrete loads,

for battery backup, or as part of a grid-connected system –
contributing energy to a utility's electrical grid during the day,
then pulling load from the grid in the evening.

PROCESS/TOOLS

In planning and sizing PV systems, Glumac utilizes Department
of Energy (DOE) software to calculate projected kilowatt hours
per year for a given location and orientation. Designers also
offer several best practices and rules of thumb when applying
photovoltaic technology:

» Influence the architecture to suit PV systems – making it
   easier and more efficient to incorporate later
» Maximize square footage if possible: regardless of
   location or orientation, the more space allocated to PV,
   the more energy produced
» Ensure consistent maintenance practices – clean panels
   will optimize output
» Understand that a 100 kW PV array, for example, does not
   always produce 100 kW of electricity. Capacity represents
   the biggest misnomer about PV systems, with energy
   production lower in early mornings and late afternoons;
   even an overcast mid-day only generates approximately
   80 percent out of each panel. As a result, designers
   recommend sizing the system's inverter so that more
   strings may be added later as needed.

FURTHER DESIGN FACTORS

The decision to pursue PV relies, first and foremost, on a
motivated client, willing to pay that higher first cost (although
rapidly falling) and dedicate an adequate percentage of available
roof façade or land area to the installed system. LEED and
government incentives continue to drive its adoption as well.
Ultimately, design for photovoltaics comes down to timing,
peak demand, and building load targets. Energy use and
storage choices, in particular, call for determining whether a
PV application will deploy net metering (and grid connectivity),
utilize batteries or operate as a stand-alone system.

The introduction of third-generation PV cells and new federal
solar initiatives make this an exciting arena as well. The
Lawrence Berkeley National Laboratory's development of a
new semiconductor material could convert nearly half of the
energy in sunlight into electricity – three times as much as most
single-layer solar cells. In addition to research on more reliable
inverters and long-duration energy storage, the DOE's *SunShot*
program aims to cut the installation of large-scale solar power to
$1 per watt without government subsidies by 2020.

DESIGN PRINCIPLE 8

USE **SOLAR THERMAL** TO BOOST HOT WATER PRODUCTION

Production of domestic hot water by converting direct sunlight into thermal energy – through use of a solar collector to absorb the heat, then transfer and store the low-, mid- or high-temperature water until needed.

Solar thermal offers a cost-effective approach to producing heating, cooling and domestic hot water for all building types. This technology continues to improve and become more popular, due in part to tax incentives and increased interest in renewable energy. Today's solar thermal technologies can meet up to 100 percent of demand – but Glumac typically recommends a hybrid system approach with back-up from alternative sources, especially during winters and nights. Components of a solar thermal scheme include solar panels, a freeze and overheat protection system, storage tanks, pumps and controls. Solar thermal production in cold weather climates must incorporate freeze protection through either a drain-back system or use of a water/glycol solution in the collector loop. Another concern is overheating, which can occur when available supply exceeds demand.

SOLAR PANELS

The process of collecting the sun's energy to produce hot water begins with a choice: flat-plate or vacuum-tube solar panel, or concentrating solar collector. The flat-plate style features copper tubing mechanically bonded to a copper sheet, which is then coated with black chrome to enhance the effectiveness of radiant energy collection. Manufacturers place the complete assembly in an aluminum frame with a glass cover to minimize conduction heat loss. Insulation on the back of the panel also helps minimize heat loss. Water circulates through the copper tubing and pulls heat off the copper plate. This style offers high heat absorption and a uniform, low profile appearance.

Alternatively, vacuum-tube panels employ a series of vacuum tubes approximately four feet (1.2 m) long, with copper plates and heat exchangers connected to a header. As the temperature of fluid inside the tube increases and vapor rises, the heat transfers to water at the top of the array, then condenses and runs down again. These panels allow

SUN + WATER + HEAT: A 3,800 square foot array powers the PV system for The Allison Inn and Spa – an 85-room luxury resort in Oregon's wine country – producing enough energy to preheat water for the facility's kitchen, laundry, guest room, pool and kitchen requirements.

SUSTAINABLE GETAWAY: Just 45 minutes from Portland, the Allison Inn and Spa offers peaceful weekend accommodations – complete with regionally-produced building materials, low VOC interiors, water conservation measures, solar hot water, high-performance lighting, and sustainable housekeeping practices.

SOLAR HEAT EXCHANGER: Critical components of the solar thermal system within The Allison include the solar heat exchanger and pump skids – one for domestic hot water and one for the resort's pool.

for higher water temperatures and dramatically minimize loss from conduction, making them ideally suited for locations with good solar radiation but cooler outdoor air temperatures. Also, they may be oriented at less-than-ideal angles and still perform quite efficiently.

Designers also recommend the use of concentrating panels when high temperatures are necessary for desalinization or for steam generation to operate a steam turbine or absorption chiller.

FREEZING AND OVERHEATING PROTECTION

To guard against the loss of panels or piping due to freezing or overheating, solar thermal systems should include one of several failsafe measures. Glumac prefers the drain-back system, specially designed to prevent both freezing and overheating. This approach also has the advantage of reducing or eliminating the use of glycol, and therefore provides better overall heat transfer efficiency. How it works: during a freeze, pumps shut off and water drains by gravity to the drain-back tank – sized to accept all water in the piping and panels on the roof. Likewise, if hot water demand has been met and the solar tank reaches its maximum temperature, the system's pumps stop and water in the piping/panels drains back to the tank.

A second freeze protection method uses a solution of glycol and water. The amount of glycol – ranging from 35 to 50

percent – depends on winter and summer design conditions. Piping and panels remain full and require no drain-back tank, resulting in a lower initial cost but also lower efficiency. The glycol system fluid must be checked and changed regularly; otherwise this mixture results in a green gooey mess, and a failed system.

To control overheating, designers take one of three approaches: utilize a large heat sink, such as a swimming pool or radiator, to divert the solar energy; add a drain-back system; or shut off circulation pumps to let solar panels stagnate. This last method, while common in the design-build solar market, offers the least desirable alternative, as it shortens the life of components and can result in hot glycol discharging from relief valves if pressures exceed allowable limits. Glumac does not recommend this approach.

PROCESS/TOOLS

Engineers at Glumac consider several key factors when designing a solar thermal system:

» *Panel array size*: Determine the optimal percentage of a building's heating or domestic hot water load based on available unshaded panel area and an economic analysis. Too many panels result in over-capacity during low loads in summer and a longer payback. Typically, solar thermal systems are most cost-effective when sized to provide approximately 50 percent of peak demand

» *Storage versus panel square footage*: Provide from 1.5 to 2 gallons of storage per square foot (3.75 kL/m3 to 5 kL/m3) of collector and insulate the storage systems

» *Pumping flow rate per collector*: Provide 1 gpm to 1.25 gpm (0.06 to 0.08 L/s) per panel – based on a typical array of six to eight panels, resulting in flows of 6 gpm (0.38L/s) to 10 gpm (0.63L/s) for each array

» *Drain-back tank size*: Provide approximately 1.35 gallons (5.1 L) per panel – based on calculating water capacity according to the roof panel and tubing

» *Heat exchanger size*: Calculate to derive the maximum energy available from panel arrays. Typically, the flow rate into and out of the heat exchanger equals the flow into the panel arrays

FURTHER DESIGN FACTORS

During the system's operation, piping from the solar collector into the building may reach as high as 300°F/148.9°C, so designs should consider thermal expansion as well. Penetrations through the roof need to be repairable, as they do not last the life of the roof itself (usually warranted up to 30 years). System designs should maximize tank storage and keep water as hot as safely possible (160 to 170°F/71.1 to 76.7°C) for mixing down and distributing to building plumbing fixtures.

Heated water can then flow from the tanks into a regular domestic hot water system. Finally, panel mounting design needs to account for panel weight, wind load, snow load and cleaning. Installing a hose bibb to facilitate cleaning of panels is recommended.

SOLAR THERMAL HOT WATER SYSTEM AT THE MIRABELLA

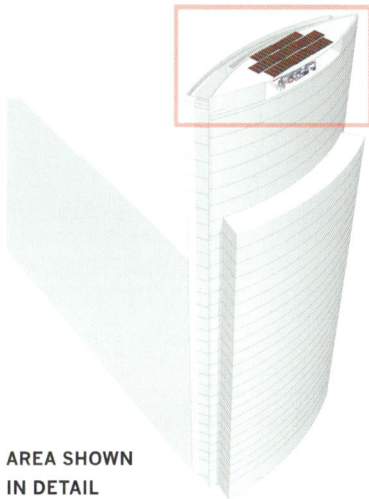

AREA SHOWN
IN DETAIL

1. Solar thermal panels
2. Hot water from panels
3. Drain back tank
4. Expansion tank
5. 15,000 gallon (56,781 l) supply tank with internal heat exchanger
6. Solar thermal collection loop pumps
7. Building hot water supply
8. Potable water supply

ROOFTOP FREEZE PROTECTION: Glumac's solar thermal design for the 30-story Mirabella Portland Retirement Community, located in the city's South Waterfront area, features a rooftop drain-back system for freeze and overheating protection. Reliance on solar hot water contributes toward an overall 45 percent return in energy cost savings for the LEED Platinum building.

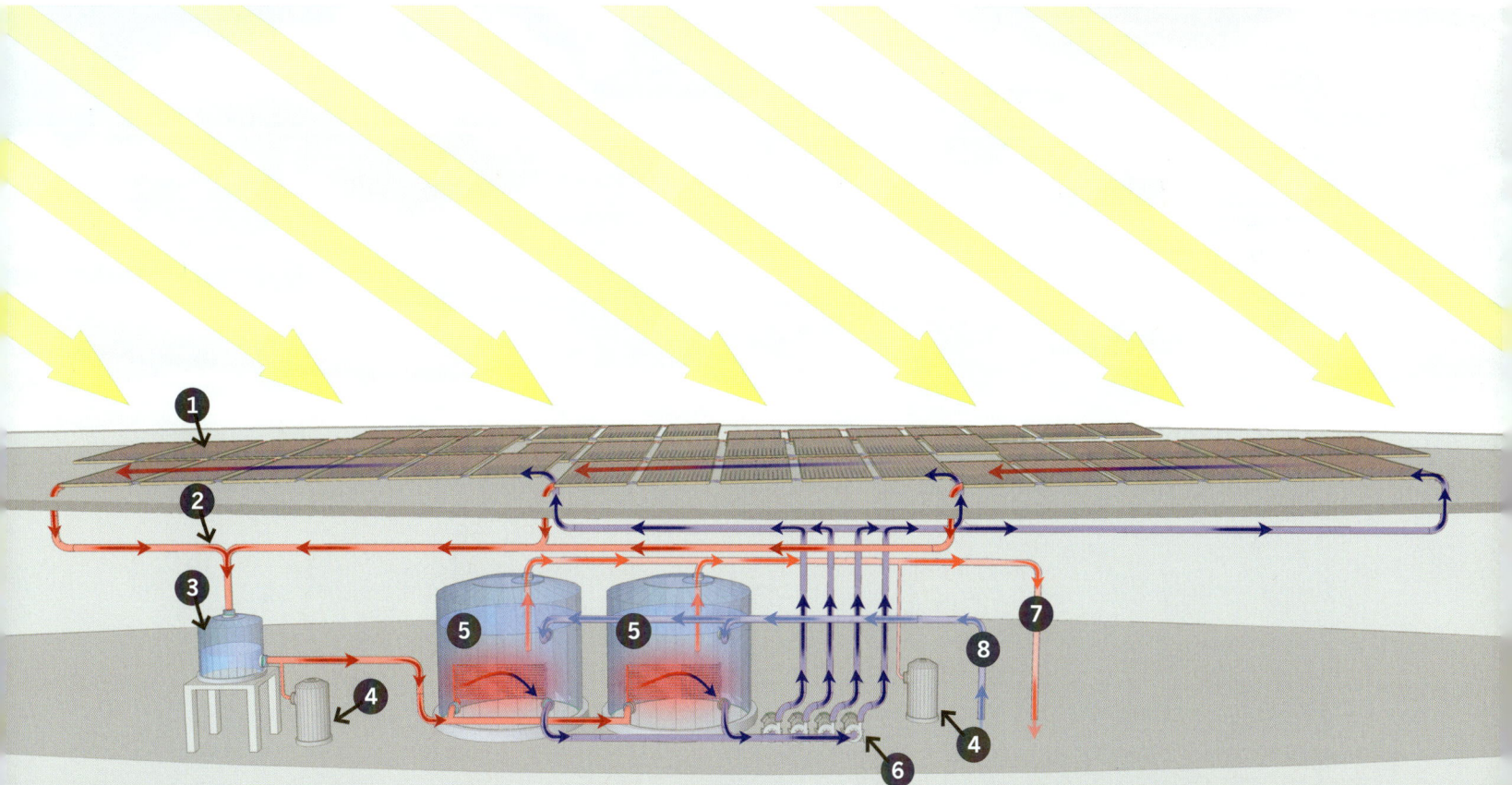

CALCULATING SOLAR POTENTIAL

The amount of solar radiation reaching a collector depends on many factors: geographic location, sun angle for any given time of the year, angle of the panel, cloud cover, pollution index, reflection from adjacent buildings, and shading from trees, hills and mountains. Based on the ASHRAE Handbook[1], the calculation for solar radiation is very complex:

$$I_{t\theta} = 128000(1+0.033412 \cdot \int \cos(2\pi \frac{N-3}{365})dN)e^{-B\int\int \sin(\sin^{-1}(\cos(L)\cos(23.45\sin(360°\cdot\frac{284+N}{365}))\cos(\Delta S\cdot 15°)+\sin(L)\sin(23.45\sin(360°\cdot\frac{284+N}{365}))))dN\cdot d\Delta S} \cdot$$

$$(\int\int\cos(\sin^{-1}(\cos(L)\cos(23.45\sin(360°\frac{284+N}{365}))\cos(\Delta S\cdot 15°)+\sin(L)\sin(23.45\sin(360°\frac{284+N}{365}))))\cdot\cos(\sin^{-1}(\frac{\cos(23.45\sin(360°\cdot\frac{284+N}{365}))\sin(\Delta S\cdot 15°)}{\cos(\sin^{-1}(\cos(L)\cos(23.45\sin(360°\frac{284+N}{365}))\cos(\Delta S\cdot 15°)+\sin(L)\sin(23.45\sin(360°\frac{284+N}{365}))))})\pm\psi) \cdot \sin(\Sigma) +$$

$$\cos(\sin^{-1}(\cos(L)\cos(23.45\sin(360°\cdot\frac{284+N}{365}))\cos(\Delta S\cdot 15°)+\sin(L)\sin(23.45\sin(360°\cdot\frac{284+N}{365}))))dN\cdot d\Delta S)$$

$$\frac{I_{dH}}{128000(1+0.033412\cdot\int\cos(2\pi\frac{N-3}{365})dN)e^{-B/\int\int\sin(\sin^{-1}(\cos(L)\cos(23.45\sin(360°\frac{284+N}{365}))\cos(\Delta S\cdot 15°)+\sin(L)\sin(23.45\sin(360°\frac{284+N}{365}))))dN\cdot d\Delta S}} \cdot$$

$$128000(1+0.033412\cdot\int\cos(2\pi\frac{N-3}{365})dN)e^{-B/\int\int\sin(\sin^{-1}(\cos(L)\cos(23.45\sin(360°\frac{284+N}{365}))\cos(\Delta S\cdot 15°)+\sin(L)\sin(23.45\sin(360°\frac{284+N}{365}))))dN\cdot d\Delta S} \cdot (\frac{1+\cos(\Sigma)}{2})$$

$$(128000(1+0.033412\cdot\int\cos(2\pi\frac{N-3}{365})dN)e^{-B/\int\int\sin(\sin^{-1}(\cos(L)\cos(23.45\sin(360°\frac{284+N}{365}))\cos(\Delta S\cdot 15°)+\sin(L)\sin(23.45\sin(360°\frac{284+N}{365}))))dN\cdot d\Delta S} +$$

$$(\frac{I_{dH}}{128000(1+0.033412\cdot\int\cos(2\pi\frac{N-3}{365})dN)e^{-B/\int\int\sin(\sin^{-1}(\cos(L)\cos(23.45\sin(360°\frac{284+N}{365}))\cos(\Delta S\cdot 15°)+\sin(L)\sin(23.45\sin(360°\frac{284+N}{365}))))dN\cdot d\Delta S}} \cdot$$

$$128000(1+0.033412\cdot\int\cos(2\pi\frac{N-3}{365})dN)e^{-B/\int\int\sin(\sin^{-1}(\cos(L)\cos(23.45\sin(360°\frac{284+N}{365}))\cos(\Delta S\cdot 15°)+\sin(L)\sin(23.45\sin(360°\frac{284+N}{365}))))dN\cdot d\Delta S} \cdot (\frac{1+\cos(\Sigma)}{2})))(\rho_g)(\frac{1+\cos(\Sigma)}{2})$$

1. Chapter 33, "Solar Energy Use," 2007 ASHRAE Handbook – HVAC Applications, Atlanta, Georgia: ASHRAE, 2007.

DEFINITIONS OF VARIABLES:

To find values for a period of time, integrals can be taken over maximum and minimum constraints for all variables and over desired constraints for time. Glumac-developed computer software simplifies these complex calculations and allows analysis and determination of monthly and annual solar capacity. Below is a summary of results for various cities at various slopes and for tracking systems.

FIXED TILT

kbtu/ft^2/yr (kWh/ft^2/yr)

ORIENTATION	SAN FRANCISCO	LAS VEGAS	SHANGHAI	LOS ANGELES	PORTLAND
Horizontal	551	658	450	574	411
South 45	620	752	451	637	458
West 45	496	573	375	514	378
East 45	455	587	377	468	341
South Vertical	400	489	269	406	310

1 AXIS TILT

kbtu/ft^2/yr (kWh/ft^2/yr)

ORIENTATION	SAN FRANCISCO	LAS VEGAS	SHANGHAI	LOS ANGELES	PORTLAND
Horizontal	728	918	516 (151)	738 (216)	528 (155)
South 45	784	992 (291)	519 (152)	789 (231)	569 (167)
West 45	608	723 (212)	421 (124)	623 (183)	454 (33)
East 45	570	736 (216)	423 (124)	579 (170)	420 (123)
South Vertical	596	764 (224)	364 (107)	593 (174)	440 (129)

2 AXIS TILT

kbtu/ft^2/yr (kWh/ft^2/yr)

ORIENTATION	SAN FRANCISCO	LAS VEGAS	SHANGHAI	LOS ANGELES	PORTLAND
2 Axis Tilt	836	1071	553	841	600

Note the variable capacity for panels at various installation angles versus the ultimate capacity of a panel that tracks the sun automatically.

MAKE THE CASE FOR
RAINWATER HARVESTING
AND REUSE

The on-site collection, treatment and storage of rainwater – typically using rooftop catchment systems – for reuse to meet a building's non-potable needs and landscape irrigation.

VESTAS RAINWATER RECLAMATION SYSTEM

1   Toilets & urinals flushed with harvested water
2   Roof top water collection with roof drains
3   Rainwater leader
4   212,000 gallon collection cistern
5   1,700-gallon (6,435 l) filtration tank (200 gallon [757 l] per hour)
6   Treatment skid
7   Pumps supply harvested water to toilets and urinals
8   Bag filter (150 gallon [568 l] per minute)
9   Filtration system loop
10 Black water discharged to sewer
11 Make-up water supply from city

RAINY DAYS: Portland, Oregon receives 208 days of rainfall annually – precipitation the new Vestas Headquarters taps through an extensive collection, diversion, treatment and storage process for reuse. As a result, this system supplies 100 percent of the building's annual irrigation requirements (64,000 gallons [242,266 l]) and 100 percent of its annual toilet and urinal requirements (539,266 gallons [2,041,344 l]) – effectively using 81 percent of the collected water.

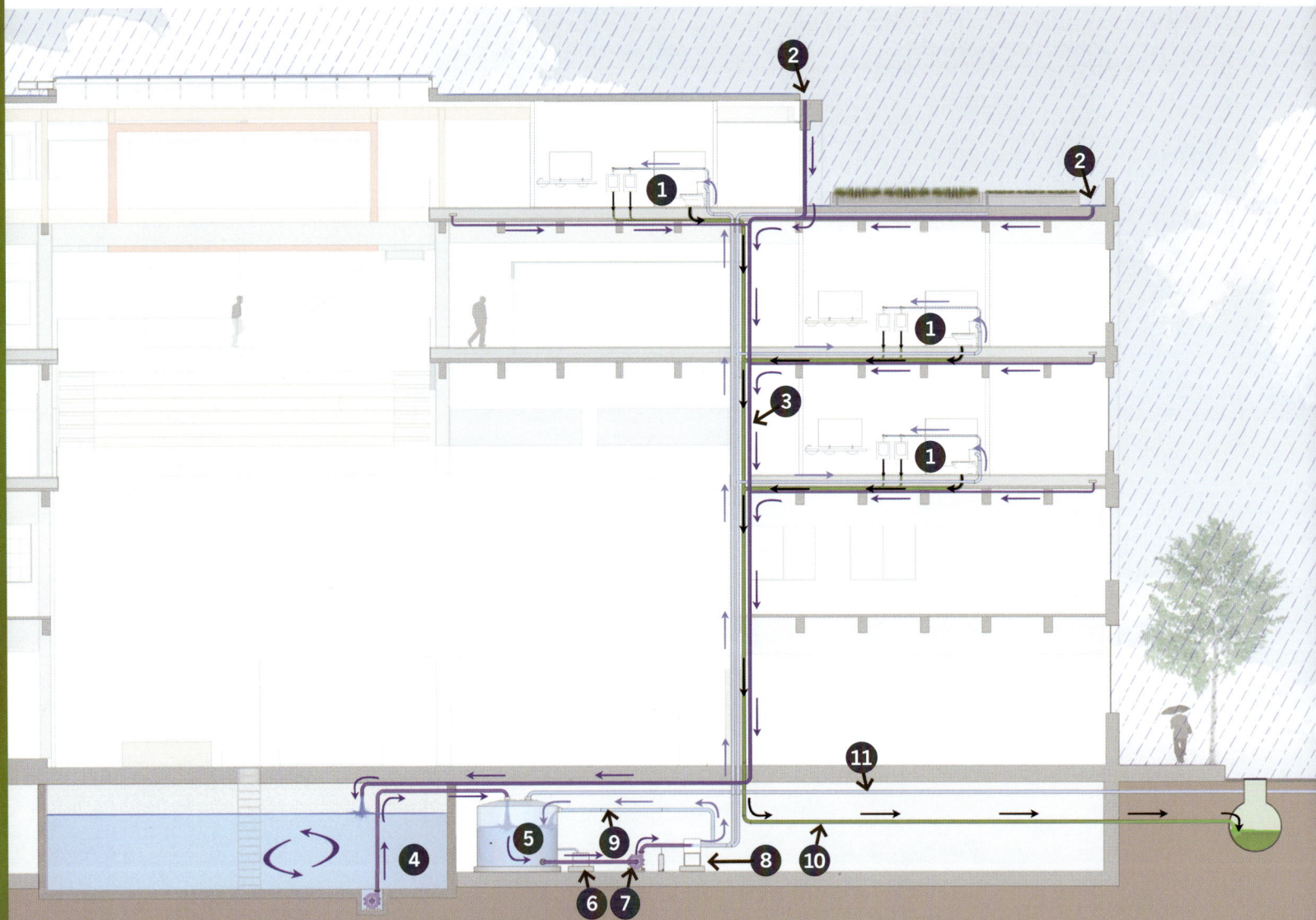

Drop by drop, water is steadily becoming North America's most precious resource. Noted for its purity and abundance in most locales, rainwater's nearly neutral pH helps prevent scale on piping and appliances while plants thrive on it. "Rainwater harvesting" signifies the collection, conveyance and storage of rain for re-use – particularly in non-potable applications such as toilets, urinals, irrigation, fire protection and cooling tower makeup. For Glumac, rainwater recovery plays an increasingly important role in its plumbing designs. At OSU's Kelley Engineering Center, efficiency measures combined with rainwater reclamation cut the building's base potable water usage by 65 percent. More recently, rainwater systems plus lower-flow lavatories and showers at Twelve|West have reduced potable water use by nearly 44 percent compared with a conventional office building; and the new Vestas headquarters in downtown Portland expects rainwater to supply up to 100 percent of its non-potable water needs.

The price of water remains artificially low across the continent, averaging between $1.50 and $2.00 for 1,000 gallons (3,785 liters) thanks to decades-long subsidies and postponed municipal infrastructure upgrades. Yet, rates also keep rising. In the last year alone, the price of water in thirty U.S. metropolitan areas increased an average of 9.4 percent, while new water quality regulations and improvements to aging wastewater treatment plants will mean continually higher sewer bills for all customers. Although water and sewer prices vary widely by region, the installed cost of rainwater recovery systems presents, at best, a long-term payback scenario.

So why consider rainwater harvesting and reuse? Glumac points to rainwater use as good stewardship and a compelling part of the green building story. Among its immediate environmental and societal benefits: extending the supply of drinking water, helping utilities reduce summer peak demand, minimizing the energy and chemicals needed to treat storm water, and reducing flow to storm water drains along with non-point source pollution. Cities with a combined sewer and water system strengthen that case further, by demonstrating how rainwater harvesting lowers demand over the life of a building. Further, payback improves dramatically when utilities provide a break in system development charges.

UPSTREAM: ELEMENTS OF THE SYSTEM

From rooftop to basement, a typical rainwater harvesting system includes seven main components. The roof serves as the primary catchment surface to collect rainfall. Glumac designers prefer clean roofing materials such as ethylene propylene diene Monomer (EPDM); however, an eco-roof functions well once the dirt leaches out after the first few months, improving the volume of rainwater recovered. Gutter and downspouts channel the rainwater to pre-filters and then the storage tank(s).

LEED and many codes require the use of pre-filters, also known as "roof washers" to remove leaves, dust and other debris from captured rainwater before diverting it to tanks. According to system demands, Glumac recommends one of several pre-filter types: a vortex fine mesh self-cleaning strainer, downspout pretreatment, or below-ground pretreatment options such as hydrodynamic separators. Next, rainwater moves into one or more storage tanks, sometimes constructed as a concrete cistern. Designers calculate the optimal tank size according to location, roof area, rainfall rate and usage (irrigation, number of fixtures, population).

DOWNSTREAM: ELEMENTS OF THE SYSTEM

Key to the delivery system, rainwater is then gravity-fed or pumped overnight to a day tank to store 24-hours worth of water – again, sized according to population, end usage and so on. A pump override keeps the tank from running completely dry. Many municipal codes also require treatment and filtration of harvested rainwater, even for non-potable uses, through a combination of ultraviolet (UV) filters, chlorinators and/or other chemicals. The UV process, for example, removes organics from water moving continuously through a closed loop system connected to the day tank. A second method integrates a chemical skid within the system, utilizing a small tank and small metering pumps that continuously treat and test water.

Other considerations in rainwater system design include make-up water controls to supply the day tank with municipal water in the event rainwater supplies fall to critically low levels. Codes also require dual piping – i.e. to separate flush water from domestic potable water serving lavatories, showers, drinking fountains and sinks. A municipal connection with appropriate backflow device is located downstream of the pump to provide water for flushing in the event of a power outage or pump failure.

PROCESS/TOOLS

Optimizing rainwater recovery depends upon two primary factors: collection potential and planned usage. Ideal for rainwater catchment would be a short, wide building with a large roof area and small population (occupancy). As an example, a building with a 10,000 square foot (929 square meters) roof area in Portland, Oregon, a city which averages 36 inches (91.44 cm) of annual precipitation, produces nearly 225,000 gallons (851,718 liters) of harvested rainwater.

Glumac engineers consider the following rules of thumb as well:

- » *Average collection rate*: 80 percent of rainfall on a conventional roof; 50 to 60 percent on an eco-roof
- » *Overall system capacity*: Use of rainwater for irrigation requires a large tank for storage over long periods (i.e. collecting and storing during rainy seasons), as well as spikes in usage thereafter; however, holding rainwater for toilets and urinals allows for a much smaller storage tank for constant use year-round
- » *Bypass mode*: If installing an eco-roof for rainwater recovery, run the system in bypass for the first few months until the storm water flows free of solids and other debris

FURTHER DESIGN FACTORS

Plumbing codes that address rainwater harvesting and reuse vary by city, county and state – as do safety measures and definitions. Even the *Uniform Plumbing Code®* (UPC), recently updated with green elements, refers only to "reclaimed water"

(tertiary treatment) rather than rainwater. However, major urban utilities – in Austin, San Antonio, Tucson – encourage rainwater harvesting as a means of conserving water, increasingly offering tax incentives and reduced storm water fees when building owners put systems in place. Some cities, like Portland, Oregon, even provide an Ecoroof Grant and Downspout Disconnect Program. Development of the new *Rainwater Catchment Design and Installation Standard*, a joint effort of the American Rainwater Catchment Systems Association (ARCSA) and the American Society of Plumbing Engineers (ASPE), also promises to advance best practices in this still-emerging field.

TEST AND VERIFY **BUILDING PERFORMANCE** THROUGH **COMMISSIONING**

Testing, validation and documentation – as a quality assurance process – to confirm that the performance of facilities, systems, and assemblies change to meet defined objectives and criteria according to the owner's operational needs.

VERIFIED!: The commissioning process – detailed, documented, comprehensive – focuses on delivering every operational element of a new building or retrofit to the owner in full working condition, from high-performance HVAC to life safety systems.

By definition, building commissioning means "proof" – validation that a facility, its systems and equipment operate as designed and constructed according to an owner's performance criteria. Commissioning also acts as a quality assurance process, offering third-party, documented confirmation focused on comfort, reliability, energy and safety.

While traditionally applied to shipbuilding, the commissioning concept took hold for the building industry in the 1980s, as recognition grew that high-performance systems were not always delivered to owners in full working condition. Such design flaws and construction defects tend to accumulate over time, leading to equipment failure, poor indoor air quality, lower productivity levels, elevated energy use and other problems. Alternatively, commissioning provides a coordinated framework to optimize building performance through design reviews, cost controls, functional testing and operator training. A recent comprehensive study by the Lawrence Berkeley National Laboratory found that commissioned buildings reduce operating costs by 8 to 20 percent compared to non-commissioned buildings (see "Commissioning By the Numbers").

NEW CONSTRUCTION: FROM START TO FINISH

Commissioning for new buildings generally follows a five-step process:

» During *Pre-Design*, the owner selects a commissioning (Cx) agent while collaborating on a preliminary scope and budget for the entire process. Glumac then conducts a workshop to develop the Owner's Project Requirements (OPR), a condensed collection of vital information about the project and, ultimately, the measure of a project's success or failure. A living document, the OPR gets revised during the project as changes are made. In addition, the Cx agent develops an initial Commissioning Plan – also refined and revised as the project progresses.

» The *Design* phase follows and involves the design team in developing a Basis of Design (BOD), which summarizes how the Owner's Project Requirements will be satisfied. Glumac reviews both the OPR and the BOD for compatibility. Next, the

THE QUALITY CHECK: Proven as a quality assurance and coordination process, commissioning spans the life cycle of a building's design, construction and occupancy – integral to any systematic approach for garnering energy savings or emissions reductions.

Cx agent thoroughly reviews the design documents against the OPR and includes Commissioning Specifications in the final construction documents.

» Upon completion of bid documents and with the construction team on board, the *Construction* phase begins. Glumac's role: to verify that building systems are installed according to the construction documents as well as functional tests to gauge their performance. By now, the Cx agent has become a familiar presence on site, conducting commissioning meetings, performing site observations and attending construction meetings.

» Once *Testing* begins, Glumac witnesses and documents all functional testing. This documentation involves recording each deficiency, then tracking to resolution to ensure there are no outstanding commissioning issues at the project's completion.

» During the *Warranty* phase, Glumac works closely with the building operating staff for the first year – fine-tuning systems and resolving any warranty issues before they expire.

COMMISSIONING FOR EXISTING BUILDINGS

Glumac also works to address the shifting operational needs and increasing energy costs of existing buildings. In *re-commissioning*, the process focuses on returning previously-commissioned buildings and component systems back to their original performance levels. *Retro-commissioning*, on the other hand, looks at enhancing operations, often in combination with control systems and VFD installations: ideal for improving, maintaining or replacing older, inefficient systems and those experiencing equipment failures. Commissioning for existing buildings typically follows a four-step procedure[1]:

» *Planning* begins with documentation of the building's current operating conditions – including an initial site walk-through, then developing the retro-commissioning plan and assembling the project team.

» During *Investigation*, the team conducts a thorough review of the facility documentation, focusing on original design documents and operation and maintenance manuals. With a full understanding of systems in place, the team performs diagnostic and functional testing according to the current operating conditions of the facility; the Cx agent then creates a "Master List of Findings" to document energy-efficient measures for the facility.

» *Implementation* involves putting the prioritized measures into action using an Implementation Plan to organize the work in a thoughtful, efficient manner that benefits the facility.

1. "Commissioning Process," Federal Energy Management Program, Energy Efficiency & Renewable Energy, U.S. Department of Energy, <http://www1.eere.energy.gov/femp/program/om_comprocess.html>

THE COMMISSIONING PROCESS

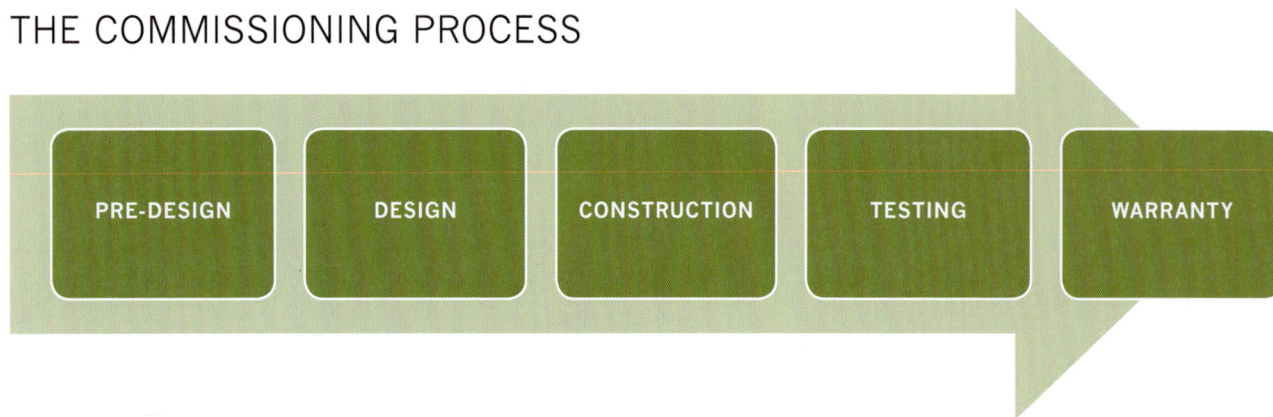

| PRE-DESIGN | DESIGN | CONSTRUCTION | TESTING | WARRANTY |

» At the *Hand-Off*, the Cx agent trains building operating staff on the proper documentation and training required to operate and maintain the new implemented measures. The project wraps up with a Final Retro-Commissioning Report, reviewed with the team during the Project Close Out Meeting.

PROCESS/TOOLS

The commissioning process relies on drawings and specifications to comprise a pre-functional installation checklist. Together, the commissioning agent and contractor document all completed tasks such as controls, piping and power before turning over the building and all of its systems to the owner. While the emphasis is on *process*, Glumac's tools of the trade include gauges and other instrumentation. In addition, agents continue to adopt smart phones and tablets loaded with scanners and cloud-based forms as more efficient, mobile alternatives for on-site assessment and documentation.

Also supporting this process, Glumac follows LEED requirements as well as *ASHRAE Guideline 0-2005,* which ensures that "a successful total building commissioning process will carefully validate interfaces and possible interferences between all building systems."[2] Additional resources include the Building Commissioning Association's (BCxA) "Best Practices in Commissioning Existing Buildings" as well as guidelines from the AABC Commissioning Group (ACG) and the American Society for Healthcare Engineering (ASHE).

ADDITIONAL CONSIDERATIONS

Ultimately, commissioning functions as a risk management tool: to detect and correct potential problems early, to reduce contractor callbacks through testing and verification, (which benefits the owner and contractor), and to provide insurance for policymakers and funders that green initiatives meet previously defined targets. Today, commissioning is a prerequisite for all newly-constructed LEED certified projects.

Another emerging discipline, *whole building commissioning*, offers great potential for new and old facilities while looking beyond the performance of energy-using equipment. Glumac commissioning agents assess the building envelope and nearly every interior detail: from windows (sealed against rainwater) to carpet and paints (for VOCs), restroom fixtures, door closures, even foundation slabs (moisture testing).

2. "ASHRAE Guideline 1-200X," Public Review Draft, Proposed Revision of Guideline 1-1996, HVAC&R Technical Requirements for The Commissioning Process, September 2006.

COMMISSIONING BY THE NUMBERS

In *Building Commissioning*,[1] developed for the California Energy Commission Public Interest Energy Research (PIER), researchers examined projects completed on 643 buildings in 26 states – representing 100 million square feet (9.29 million square meters) of floor space and $43 million in commissioning expenditures. The 2009 report noted its results as "compelling," revealing "over 10,000 energy-related problems, resulting in 16 percent median whole-building energy savings in existing buildings and 13 percent in new construction, with payback time of 1.1 years and 4.2 years, respectively."

The median normalized cost to deliver commissioning stood at $0.30 per square foot for existing buildings and $1.16 per square foot for new construction (or 0.4 percent of the overall construction cost). Meanwhile, commissioning for high-tech buildings tended to achieve greater cost effectiveness and energy savings.

Finally, the study summarized:

"Further enhancing the value of commissioning, its non-energy benefits surpass those of most other energy-management practices. Significant first-cost savings (e.g., through right-sizing of heating and cooling equipment) routinely offset at least a portion of commissioning costs – fully in some cases. When accounting for these benefits, the net median commissioning project cost was reduced by 49 percent on average, while in many cases they exceeded the direct value of the energy savings. Commissioning also improves worker comfort, mitigates indoor air quality problems, increases the competence of in-house staff, plus a host of other non-energy benefits."

1. Mills, Evan Ph.D. *Building Commissioning: A Golden Opportunity for Reducing Energy Costs and Greenhouse Gas Emissions*, Berkeley, California: Lawrence Berkeley National Laboratory, July 21, 2009. <http://cx.lbl.gov/2009-assessment.html>

PART III

THE PROJECTS

GERDING THEATER, 2006: Home to Portland Center Stage, the Gerding Theater at the Armory attracts more than 150,000 visitors for Off Broadway shows, a Shakespeare festival and hundreds of performances and community events each year.

CHAPTER FIVE:

GERDING THEATER AT THE ARMORY

A LITTLE NIGHT MUSIC

"The stage is not merely the meeting place of all the arts, but is also the return of art to life."

— *Oscar Wilde*

"The Main Stage is the center of the building, really the heart of the project. The design started here to ensure that we would have optimal performance space and yet be scaled small enough for actors: to naturally project their voices, with optimal sight lines, visual and acoustic properties, good lighting and sound. People really love the space – all the actors, directors, designers who come here. For us, it's mostly about what the patrons think – and that has really paid off."

— Creon Thorne,
General Manager, Portland Center Stage

PORTLAND, OREGON

The President spoke here. It has been a venue for operas and symphonies, political rallies, tennis matches, dog shows, boxing matches, and Portland's first moving picture show. Originally constructed in 1891 to house local units of the Oregon National Guard, the Romanesque Revival building also reigned as the largest assembly space in the city until the big movie palaces were erected in the 1920s and 1930s – and it later hosted auto shows and even televised championship wrestling.

Still functioning as a Armory through most of the 20th century, by 1969 the building was sold to the adjacent Henry Weinhard Brewery, serving as a beer warehouse for more than thirty years. For its castle-like turrets, parapets and narrow gun-sight windows, the Portland Armory had long been one of the neighborhood's defining elements. Yet, by the late 1990s, the building was in dire need of repair and eventually declared a fire hazard.

In 2000, developers Gerding Edlen purchased the First Regiment Armory as part of a five-block downtown parcel surrounding the historic Blitz-Weinhard Brewery. Over the next several years, the "Brewery Blocks" would become a successful mixed-use urban development of shops, condos and offices within the city's fashionable Pearl District. Following completion of its first project, a 15-story condominium on the same block as the Armory, the developers decided against tearing down this local landmark to construct another new building. Instead, Bob Gerding wanted to find a single tenant for the space. Retail stores, restaurants, a fitness club – all considered the Armory, but the huge, single-room structure was ill-suited for redevelopment. In addition to project funding, the building faced other challenges as well, such as strict preservation guidelines and the need for seismic upgrades. This historic structure seemed destined for demolition.

THE OVERTURE: The entrance lobby welcomes theater-goers to the restored 19th Century Armory, a fortress-like space that highlights fresh air and daylight, with a concrete thermal mass for cooling and a dramatic lighting display high overhead.

"We had to fit a modern building inside an historic shape, complicated by a structural box in the middle – the theater – so everything had to run down these 'highways.' It was a challenge to fit modern MEP systems within these structural constraints and architecturally make it look like there's nothing there. So we used newer technologies like underfloor air, displacement ventilation and chilled beams to fill the needs. And get Platinum at the end of the day."

— *Bob Schroeder, Principal, Glumac*

Then Portland Center Stage (PCS) stepped forward. The city's largest theater company sought a new home, a more distinct, intimate venue than their leased space in the City Performing Arts center twelve blocks away. Together with Gerding Edlen and the Portland Family of Funds, PCS formed a unique public/private partnership to launch a $28 million project for the Armory's renovation. The result? A state-of-the-art performance hall featuring a 600-seat main stage, 200-seat studio theater, offices, rehearsal space, gallery and café.

Named in honor of Bob and Diana Gerding, the new theater opened to sellout crowds in October 2006 and now attracts more than 150,000 people annually. The Armory achieved yet another milestone as the first building on the National Register of Historic Places and the first performing arts center in the U.S. to achieve LEED® Platinum certification. Filled with light and fresh air, the building's green debut features skylights, chilled beams for heating and supplement cooling, displacement ventilation throughout, and a high-efficiency lighting design that evokes a theatrical experience. In 2007, the project was awarded an Honorable Mention as one of the AIA Committee on the Environment's Top Ten Green winners.

ENVIRONMENTAL OBJECTIVES

The Pearl District has quickly gained renown as a pioneering area in terms of green urban design, mass transit and historic preservation. So it came as no surprise that Gerding Edlen intended to design all six buildings in the Brewery Blocks to at least LEED Silver standards and, as its signature project, mandated that the theater renovation must achieve LEED Platinum.

Environmental responsibility was an essential driver of the building design as well as the theater company's relationship to the community. Like many recent construction projects downtown, the renovation needed to be sensitive to the nearby Willamette River's water quality, which is regularly threatened during large storms due to overflows from the city's combined sewer system. Maintaining the historic character of the building was critical as well. As a result, the team focused on reuse of materials to conserve the embodied energy of the existing brick, stone and wood trusses while also preserving the craftsmanship and aesthetics of the original 30-inch-thick (76.2 cm) brick walls and rusticated basalt stone foundation.

DESIGN OVERVIEW

With the goal of LEED Platinum established early, the design team was tasked with finding solutions that met sustainability goals while preserving the building's architectural integrity. As the last of six major buildings to be completed in the Brewery Blocks redevelopment, the Gerding Theater project benefited from an integrated design team with a great deal of experience working together. In addition to Glumac and GBD Architects, the team included Green Building Services as LEED consultant and eco-charrette lead.

Faced with ambitious goals for building performance while also working within the constraints of the site, budget and historic-tax-credit guidelines, the planning process required more iterations – and more design time – than usual. To meet specific LEED requirements, many systems were still being designed late in the construction documents phase.

Unquestionably, the design team's biggest initial challenge was how to fit a fully-scaled theater plus offices and rehearsal spaces within the old building, something many of them likened to "building a ship in a bottle." PCS requested at least 55,000-square feet (51,000-square meters) of program space, whereas the Armory's footprint measured out at just 20,000-square feet (19,000-square meters). Theater design consultants recommended punching a hole through the roof to install a full-height fly tower rigging system that would have interfered with the building's historic roof shape.

Instead, the team opted to treat the interior theater as a self-contained building. This choice meant digging thirty feet below street level to keep the building shell intact while creating enough volume for the main stage, a "black-box" theater (placed underground), rehearsal spaces, and administrative offices. Excavating near the adjacent high-rise required careful shoring techniques to protect the building's original foundations and 30-inch-thick (76.2 cm) brick-bearing walls. This approach allowed the team to retrofit the half-block-sized floor plate with a massive concrete box inside the existing shell; this design, in turn, served to seismically brace the structure and acoustically isolate performance and rehearsal spaces.

As a result, the Armory renovation retained all exterior brick walls in their entirety, leaving the interior masonry walls exposed along with 200 by 100 feet (61 by 30.5 meters) of column-free space that features parallel-chord bowstring trusses with tie rods at the bottom. Ultimately, 79 percent of the original structure was reused, with more than 95 percent of construction and demolition debris recycled.

The project also features minimal use of finish materials to conserve resources and reinforce the character of the original building. Over 58 percent of the wood used for millwork, finish carpentry, doors, and formwork is Forest Stewardship Council (FSC) certified. GBD Architects selected low VOC paints, adhesives and carpets, as well as composite woods free of urea-formaldehyde. The Structural steel contains 90 percent recycled content.

ROMANESQUE REVIVAL: Constructed in 1891, it has been a performance hall, the site of exhibitions and auto shows, and home to the Oregon National Guard.

Armory, Oregon National Guards, PORTLAND, Oregon.

Tony: [singing] There's a place for us, Somewhere a place for us.

— *West Side Story*,
Gerding Theater's opening night performance, October 1, 2006

Many complex design decisions were resolved collaboratively among the various disciplines. For example, concrete (made up of 10 to 40 percent fly ash) floors helped to provide structural support so that additional steel bracing was not needed, thereby keeping the building's brick walls unobstructed and its massive old-growth-timber ceiling trusses viewable from the multi-story lobby. The constraints of the existing shell forced the design team to make efficient use of the available space. Placing the rehearsal studio and staff kitchen with the top floor administration offices allowed the kitchen to double as a lounge and informal meeting space. Likewise, the excavated lower floors – including the Studio Theater, dressing rooms, actors' lounge and storage – were placed adjacent to compact spaces designated for the building's air handlers, gas boilers, hot water tanks, ventilation intake and outtake, and automation control systems.

Rounding out the program above are several public spaces, including the two-level lobby and outdoor park available for classes, lectures, special events, outdoor performances and festivals.

MEP DESIGN PROCESS

Beyond the building's reuse and emphasis on recycled materials, the MEP design strategies were the biggest contributors to achieving LEED Platinum certification. Since the building massing, envelope, orientation and footprint were already established, Glumac focused on blending a mix of new green technologies with the existing historic structure to meet LEED standards for healthy indoor environments.

"We recognized the actors, crew, stage managers and staff all spend a lot of hours in this building," said Bob Schroeder, Glumac's project manager. "So we wanted to give them a building that they felt comfortable with, that met their needs and was a pleasant place to be."

Engineers crafted a comprehensive HVAC approach based on energy and computational fluid dynamics (CFD) modeling performed by Green Building Services. Glumac's design scheme included displacement ventilation, radiant flooring and occupant-controlled chilled beams to efficiently regulate temperature and air quality on a highly localized basis. Skylights, hidden behind a parapet on the roof, allow natural light into administrative areas and the double-height lobby. Advanced glazing maximizes daylighting while minimizing winter heat loss and summer heat gain. A rainwater catchment system, also on the historic roof, delivers gray water to flush toilets and urinals.

The mild Northwest climate also allowed project designers to leave the inner face of the brick shell exposed without compromising energy performance. This existing mass, along

THE MAIN THEME: The 600-seat Main
Stage theater lies at the heart of the
building, utilizing displacement ventilation
underneath to create perfect conditions for
the audience during every performance.

BLEND OF OLD AND NEW: The building's renovation, architectural design elements and MEP systems successfully combine new technology with the existing historic structure – from displacement ventilation in the lobby to underfloor air, chilled beams and integrated light fixtures within offices on the mezzanine.

with the new concrete floors and walls, serves as a thermal "flywheel" to reduce diurnal temperature swings. Other energy-saving features include daylight controls and occupancy sensors. In addition, the building is connected to an efficient district-chilled-water plant.

INDOOR AIR

Like most performing arts spaces, the Gerding Theater is often unoccupied for much of the day but requires high rates of ventilation – with little noise – and cooling loads during the few hours of a performance. Rather than introducing air overhead – among the catwalks and lighting loads – at higher velocities, Glumac recommended a displacement ventilation system for the main theater. This approach delivers conditioned air under every other theater seat, providing greater uniform temperature and ventilation throughout the audience: air naturally rises as it warms, drawing in more supply. In this case, the concrete box created within the Armory shell offered the perfect underfloor air plenum. Noise was reduced through the use of a fan wall within the air-handling units in mechanical rooms, made up of multiple smaller fans not requiring sound-trap elements in the ductwork design. Electronic filtration also improves air quality.

In addition, displacement and underfloor ventilation were installed in the theater lobby. Underfloor air distribution in the building's top-floor work spaces provides individual control of floor vents with adjustable lighting and cooling from an integrated chilled beam/light fixture system above.

Carbon dioxide (CO_2) sensors in air handling units allow for demand-based ventilation in all occupied spaces. Mechanical systems are zoned to deliver appropriate amounts of outside air and maintain comfort in offices, theaters, rehearsal halls, the lobby and any other gathering spaces that experience big swings in population.

WATER & WASTE

Water efficiency measures within the theater include low-flow fixtures and dual-flush toilets. Among the project highlights is a rainwater harvesting system on the roof, where water is collected, filtered and diverted to flush the toilets and urinals. Through a combination of storm water swales, limited parking areas and native plantings, this system reduces runoff entering the municipal sewer system by an estimated 26 percent, compared with a conventional system.

Overflow from the system's 12,000-gallon (45,420-liter) storage tank and a portion of the sidewalk runoff drains to a park alongside the building, featuring native plantings and pervious pavers.

HVAC AND DAYLIGHTING

1 Skylights

2 Radiant chilled beams

3. Fresh air intake

4 Radiant heating and cooling slabs (shown in cooling mode)

5 Fan wall air handler

6 Supply air

7 Underfloor air distribution system

8 Displacement floor diffusers under theater seats

9 Theatrical lighting creates heat that drives the displacement ventilation's convection currents

10 Return air duct

11 Exhaust air duct

12 Chilled water supplied from district plant

13 Chilled water pump supplies water for radiant cooling elements

14 Water heater for radiant heating slab

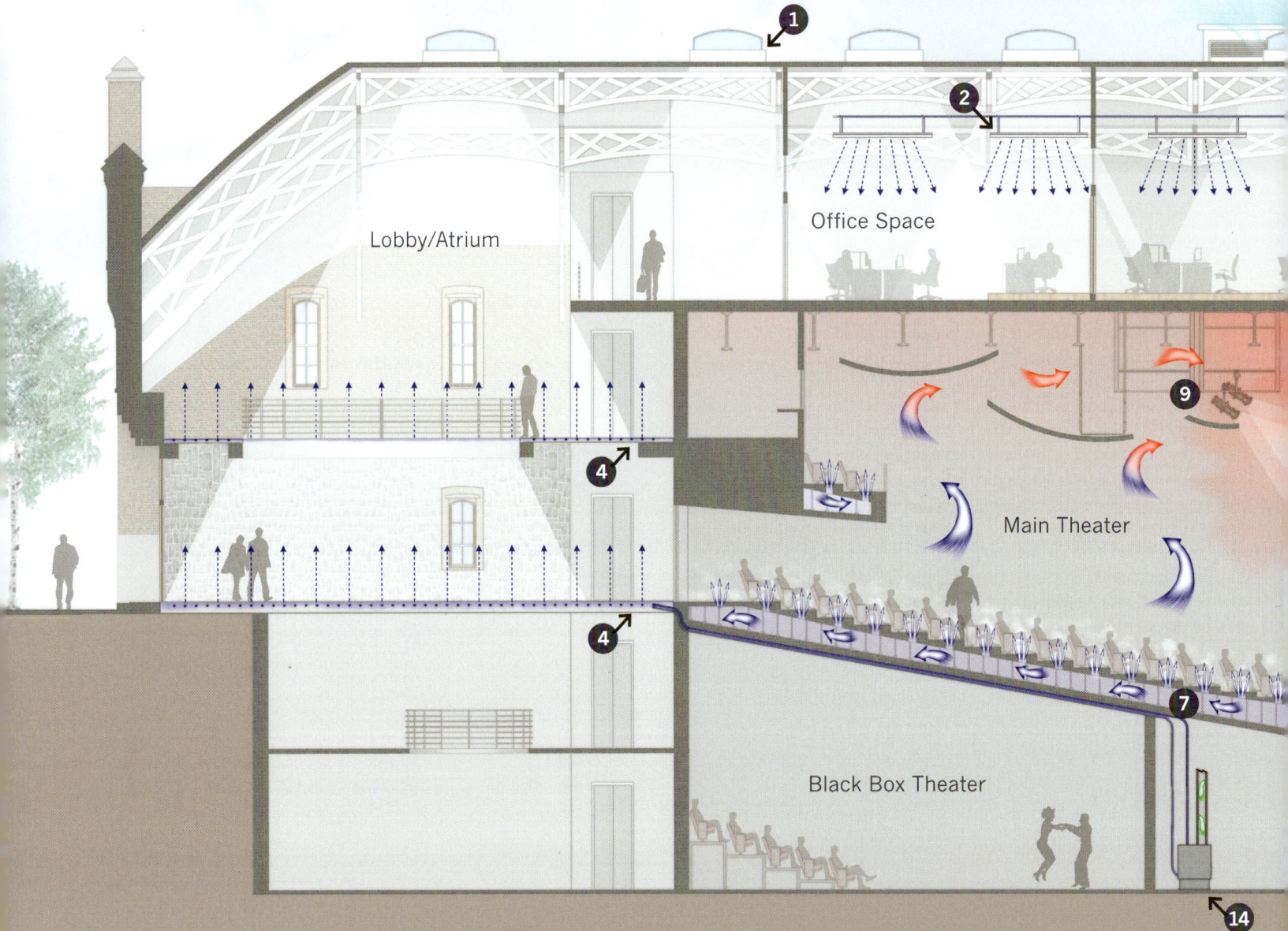

Office Space

Lobby/Atrium

Main Theater

Black Box Theater

WATER, AIR, LIGHT: While preserving the historic integrity of the existing structural shell (brick, masonry, bow trusses), the Theater's new design takes full advantage of both active and passive strategies to condition varied spaces throughout the building.

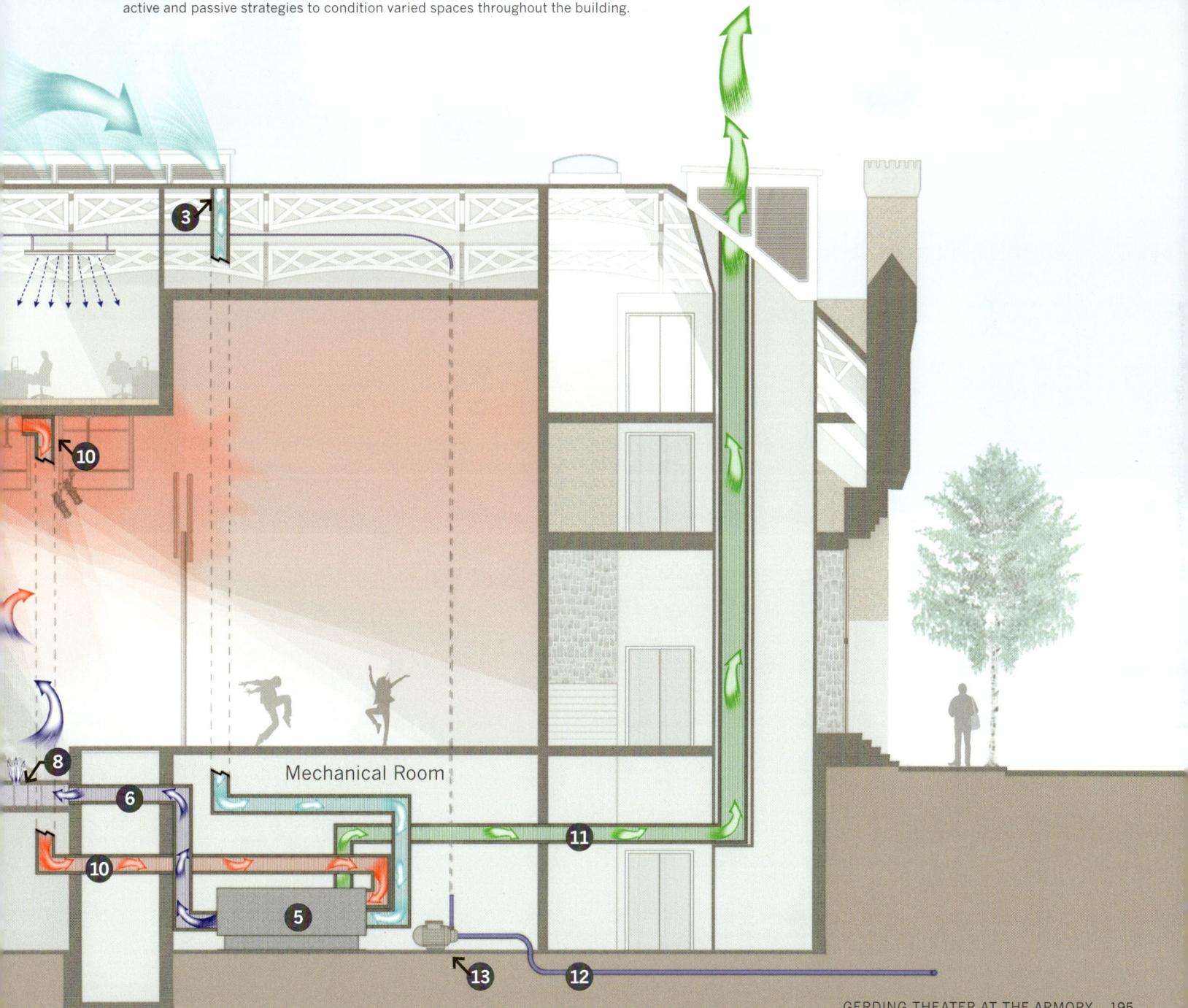

3

10

8

6

10

Mechanical Room

5

11

13

12

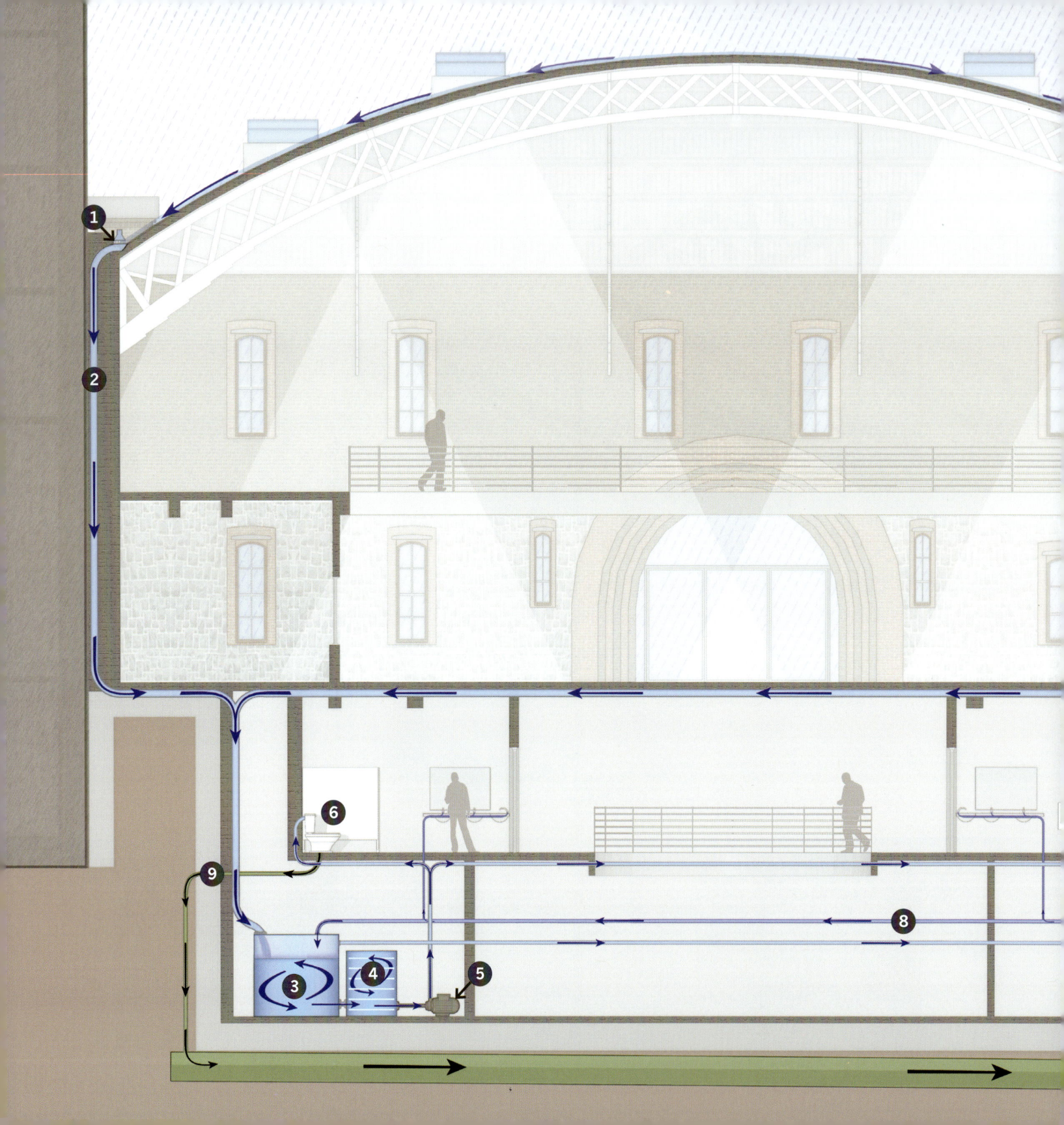

RAINWATER HARVESTING

Driven by measures to protect water quality for the nearby Willamette River, the building collects and filters rainwater for use in the building. The system reduces potable water use by an estimated 40 percent and storm water runoff by 26 percent.

1 Perimeter roof drains collect rainwater
2 Downspouts take water from drains to an underground cistern
3 The 12,000-gallon (45,420-liter) cistern holds harvested rainwater for use in toilets and urinals
4 Ultraviolet filtering system treats water before it is pumped to restroom fixtures
5 Pumps supply reclaimed water to toilets and urinals
6 Low flow urinals and dual flush toilets use harvested water, reducing demand on the municipal water supply
7 Rain garden/bio-swales filter and retain sidewalk runoff and use harvested rainwater overflow when available for irrigation
8 Municipal potable water supplies sinks in restrooms and tops off the cistern during dry months to keep supply constant
9 Wastewater and black water flows out to the municipal sewer system

"Whenever you have a very confined space with aggressive architectural programs and finishes and intensive mechanical/electrical/plumbing systems, there's going to be conflicts. It was very encouraging to see the way all the team members – the architects, all the trades – played together."

— *Kirk Davis, Principal, Glumac*

HEATING & COOLING

Passive chilling (due to the brick shell and inner concrete structure) and air-circulation features throughout the building combine to reduce the project's anticipated energy use for mechanical systems by 40 percent. Special glazings in the skylights and windows maximize daylighting while controlling passive solar heating and minimizing seasonal heat losses and gains. Additionally, hot water can be circulated to provide warm air via a small fan. High-efficiency, gas-fired condensing boilers provide the remainder of the building's heating needs.

Glumac designed the Theater's heating and cooling system to work in concert with the underfloor ventilation system. Displacement ventilation in the lobby, theaters and offices relies on the buoyancy of warm air to draw stale air up and away from the occupants' breathing zone. Engineers also recommended that "active" chilled beams be placed strategically on several floors – one of the first applications of its kind in Portland. Particularly because administrative work areas have limited ceiling heights, the chilled beams supplement both cooling and heating, providing more efficient, occupant-controlled comfort than a conventional fan-driven mechanical system. These overhead suspended fixtures (with integrated lighting) heat and cool the space around a person, allowing air cooled by coming in contact with radiator fins to sink at low velocity to the occupied zones below.

And finally, the theater is tied into a district chiller plant, contributing further to the building's energy efficiency. Located nearby, the plant supplies low-temperature chilled water to all five of the Brewery Blocks through an underground garage.

LIGHTING & DAYLIGHTING

Introducing daylight into the Armory posed an early design challenge, since the existing shell had only small windows and gun slots. Due to preservation restrictions, new windows were not allowed, and any new skylights could not be visible from the street. Instead the project team chose to strategically position forty-two skylights, seventeen of them operable, on the north side of the roof; as a result, the top of the theater volume – including offices, the rehearsal hall and the entrance lobby – now receives natural light and fresh air. Approximately 75 percent of regularly-occupied spaces are daylit, which represents roughly 25 percent of the total building.

Dimmable compact and linear fluorescent fixtures supply the remaining general illumination, except for the lobby halogens. Lighting control strategies rely on "open-loop" photoelectric daylight sensors along with "closed-loop" sensors (occupancy), which measure reflected light levels coming off the work plane. In the open offices, a semi-direct pendant fixture with T5 and T5 HO lamps bounces 20 percent of its illumination off the trusswork and ceilings; the rest spills down through a blade louver.

THE REPRISE: Beyond performances, the Gerding Theater celebrates art through interactive exhibits, education and outreach programs, receptions, readings and many other public events.

SITE

As originally built, the Armory fully occupies a 20,000-square foot (19,000-square meter) lot and shares the block with a 15-story residential tower. A "sliver park" along the long edge of the building offers outdoor seating and native vegetation. Pervious pavers increase rainwater infiltration on site, while street trees and a high-emissivity roof reduce the project's contribution to the urban heat-island effect.

ENERGY

A condominium constructed directly to the south shades the Armory, limiting solar access and making passive solar design or on-site power generation nearly impossible. However, the mix of HVAC technologies within the building are tied together by an energy management system. In addition, PCS purchases renewable energy for all of its electricity needs.

COMMISSIONING

In-depth commissioning was carried out by an independent team at Glumac to ensure that the building's mechanical, lighting and filtration systems would operate as designed. With the theater's opening night fast approaching, the window to correct any issues was much shorter than normal. Over the course of a week, during dress rehearsals and while dealing with flushout issues, engineers completed the work in time for Portland Center Stage's first performance in its new home.

"Wieden+Kennedy is all about attracting the best creative talent in the world – and part of that is having this super energy-efficient work space with fresh air supplied through a raised access floor, daylighting, and a design that balances comfort and life safety. We used technology to help the architect achieve what I think is spectacular architecture."

— *Steven Straus, President, Glumac*

WIEDEN+KENNEDY

CORPORATE HEADQUARTERS
PORTLAND, OREGON

The energy here, both creative and in terms of comfort, all quite literally stems from the high-rise atrium at the heart of this advertising agency's Portland home. Named AdWeek's "Global Agency of the Year" in 2007, Wieden+Kennedy came to prominence in the 1980s with its iconic work for Nike and the tagline "Just Do It." By 1999, converting a former cold storage warehouse (built in 1910) into a multi-level office and retail space became the latest in a long line of inspired ideas by the firm. Designed as a sustainable workplace to foster creativity and collaboration, W+K's corporate headquarters features an underfloor air distribution system, daylighting of interior office spaces through the atrium, operable windows, exposed thermal mass, and an economizer capability on HVAC systems.

"...incorporating spaces for the creative arts. Designed with these aims in mind, the building speaks strongly of new values."

— Architectural Record, *October 2001*

PROJECT DETAILS

Six-story, 220,000-square foot (20,400-square meter) office space, 32,000-square foot (2,970-square meter) retail space, 8,000+-square-foot (2,970-square meter) central atrium

Completed December 1999

Architect: Allied Works Architecture

Contractor: R&H Construction

Developer: Gerding Edlen

Awards: 2000 Portland's BEST Award, ASHRAE Regional Technology Award

"At The Casey, our role was as an advisor to the owner, putting together initial sustainable design concepts and working with subcontractors to actually implement those ideas. So I'm proud of Glumac's accomplishment in contributing to this groundbreaking effort, which in many ways also anticipated the development of an integrated project delivery method."

— *James Thomas, Vice President, Glumac*

THE CASEY

Active, contemporary and luxurious, The Casey represents yet another sustainable jewel within downtown Portland's revitalized Pearl District. Designed to use 52 percent less energy and 32 percent less water than a conventional building of its type, the 16-story residential tower features solar panels to generate on-site power and an eco-roof for cooling and storm water management. Also contributing to performance: high-efficiency glazing and curtain-wall design, ground-source heating and cooling, and operable windows were placed throughout the residential and retail areas. In addition, every condo unit contains an innovative heat-recovery device which uses exhaust from the kitchen and bathroom to pre-heat fresh air that is introduced mechanically back inside – making a significant difference in building energy efficiency.

"In addition to being the first high-rise condominium in the country to achieve LEED Platinum certification, The Casey represents a partnership between the building's developers, designers, and the local arts community."

— *Kristin Dispenza, "The Casey: A High-Rise Condominium Earns LEED Platinum," Green Building Elements, May 20, 2008*

PROJECT DETAILS

Sixteen-story, 159,646-square foot (14,832-square meter) high-rise with sixty-one residential condominiums and ground-level retail; four levels of below-ground parking (46,900-square feet or 4,360-square meters)

Completed November 2007

Architect: GBD Architects

Contractor: Hoffman Construction

Developer: Gerding Edlen

LEED Platinum certified

KELLEY ENGINEERING
CENTER, 2005: The
first LEED Gold research
building on the West Coast,
the Center ushered in a
new era for OSU's main
campus – promoting the
possibilities of electrical
engineering and computer
science as part of a vital
sustainable design future.

CHAPTER SIX

KELLEY ENGINEERING CENTER, OREGON STATE UNIVERSITY

IN THE HALLS OF HIGHER LEARNING

"Being digital is different. We are not waiting on any invention. It is here. It is now. The control bits of that digital future are more than ever before in the hands of the young."

— *Nicholas Negroponte, from* Being Digital

"Kelley's atrium is a beautiful space, but mostly I like how it functions: it's very efficient and central to the natural ventilation system of the building. While inside, you and other people are creating the heat column that makes air rise to draw fresh air through the rest of the building. It's always busy, always alive in there."

— John Gremmels
Senior Project Manager, *Oregon State University*

CORVALLIS, OREGON

Its airy, light-filled atrium is the perfect place to study "Geometric Modeling," "RF Circuit Design" and countless other technical subjects. Or to collaborate with fellow students, researchers, visiting scholars and faculty.

Located on the north side of campus, the Kelley Engineering Center is home to the rapidly growing School of Electrical Engineering & Computer Science (EECS). From sky bridges and hallway alcoves to glass-walled conference rooms and graduate student offices clustered around research laboratories, the 153,000-square foot (14,200-square meter) building provides classrooms and work spaces for over one hundred-fifty faculty members and three hundred graduate students. As designed and built, the Center signifies a dramatic departure from most academic engineering buildings: its soaring, transparent atrium provides ample spaces for interaction and collaboration, complete with a café and open study areas. This central space also serves as a mechanism to bring daylight and natural ventilation to interior labs and offices, reducing energy costs while achieving a comfortable academic teaching and research environment that fully communicates sustainability.

Glumac provided sustainable engineering design services, teaming with Yost Grube Hall Architects (YGH) and Skanska USA to achieve the first LEED Gold academic engineering building on the West Coast.

The four-story building also represents the centerpiece of an ongoing push by Oregon State University (OSU) to become one of the top twenty-five engineering schools in the U.S. The university's College of Engineering understood that one path to a top-tier program would be attracting distinguished professors and outstanding students, while boosting its research to retain existing talent within the school. Enter Martin N. Kelley, a 1950 OSU civil engineering graduate. His $20 million donation (along with $20 million in matching funds from the State of Oregon and $5 million in private gifts) served as the

WORLD-CLASS RESEARCH: While creating an environment for learning and discovery, the university also envisions the 153,000-square foot building (14,214-square meter) as a focal point for advanced research in OSU's drive to establish one of the nation's top-25 engineering programs.

ENHANCED COLLABORATION:
The Center's signature
four-story atrium, featuring
walkways to connect each
floor and spacious public
gathering areas, was
designed to bring faculty,
students and visitors
together in a setting that
encourages communication
and collaboration.

"This LEED Gold certified building is projected to be 43-percent more energy efficient than code requires. And, perhaps of equal importance, it is a habitat that teaches, through its design, the future professionals who will be responsible for further improving our standards of sustainability."

— *"Oregon Engineering,"* ArchitectureWeek, *p. E1.1, 15 November 2006*

catalyst for construction of a new OSU engineering center that would bear his name.

It was established early that the Center would emphasize shared resources, places to foster communication among multidisciplinary teams, and opportunities for dialogue and to brainstorm new ideas that translate into cutting-edge research. Upon the project's launch, Ron Adams, Dean of Engineering at OSU, noted: "Today, innovation is all about collaboration, teamwork, and new ideas. This new building is designed to help spark those ideas by ensuring that the people inside connect. Because out of these connections comes collaboration, and that is the key underpinning of innovation."[1]

Since its grand opening on October 29, 2005, the Kelley Engineering Center has become an effective recruiting tool to attract top talent and home to many high-profile research projects, such as a private start-up software firm where OSU faculty and students work directly with the company's executives. Other successful research collaborations include development of the world's first transparent electronics and OSU electrical engineers and avian ecologists working together to track the migration routes of songbirds using miniature cell phones.

1. "Engineering Center Designed For Collaboration, Sustainability," Gregg Kleiner, OSU

ENVIRONMENTAL OBJECTIVES
Beyond addressing critical space needs with a modern facility that would drive world-class engineering research, the University mandated that the forty-five million dollar project would meet high green building standards with at least a LEED Silver certification. The new Center was also required to highlight advanced systems – efficient climate control, standby and UPS power, and wi-fi systems – that were understood and appreciated by occupants. Finally, the design had to respect the historic character of the OSU campus.

Together, the University and design team realized that achieving LEED Gold certification was easily within reach of project goals. Given Oregon's ideal climate for natural ventilation, one key goal was to bring outside air into as many interior spaces as possible. Designers also planned to maximize daylight in all areas not containing sensitive equipment. These measures, combined with several more sustainable design strategies, could result in using 50 percent less energy than comparable science centers – making it the greenest academic engineering building in the nation.

DESIGN OVERVIEW
After working together on renovation of OSU's Dixon Recreation Center, an indoor practice facility on campus, Glumac and YGH interviewed for and won the contract to design the Kelley Engineering Center. At Dixon, they had successfully incorporated several innovative HVAC technologies within a very constrained

"It's the most sophisticated natural ventilation scheme we've ever done; in the past, we have naturally ventilated a lobby and perimeter offices and an atrium space, but never a whole building. The same for daylighting. So Kelley has the ability to be 100 percent naturally lit and 100 percent naturally ventilated."

— *Steven Straus, President, Glumac*

budget, including natural ventilation for all gymnasiums and the aerobics room, as well as heat reclaim for the pool.

This new engineering building project was launched with a series of eco-charrettes. As part of the project team, Glumac spearheaded all mechanical, electrical and plumbing design. Green Building Services[2] was brought in as sustainability consultant. Participants then identified a range of potential green strategies to consider. Despite the team's early concerns about broad acceptance by the OSU facilities group, the integrated design process proved successful in developing cost-effective building concepts to reduce energy, including prominent use of daylight and natural ventilation throughout the building.

YGH initially developed a master plan for the engineering center on campus. The new building would be the first of several buildings totaling 500,000-square feet (46,500-square meters) of engineering space. Located along Engineering Row and adjacent to OSU's College of Business, the Center's design had to complement and enhance its historic setting within the campus core. In massing and materials, the architecture responds to nearby buildings: featuring exteriors of red brick, limestone and pre-cast trim, recessed windows and entry portals framed in stone along with modern detailing and the expansive glass-paned atrium. These materials were likewise carried inside and combined with exposed concrete, steel, aluminum panels and wood to create a warm, active interior.

Bisecting the building is the 74-foot-high (22.5-meter) atrium, its roof designed to diffuse and soften the southern light. The atrium's north and south sides feature stretches of glass allowing visitors to view projects in adjacent graduate research assistant (GRA) work areas on the second through fourth levels, while offices for faculty are located along the Center's outside perimeter. Laboratory spaces are located further south between the GRA and faculty areas, an arrangement designed to foster communication and allow natural light throughout the offices. Across four floors are wireless classrooms, office clusters, and common areas to encourage communication. These common areas include "plug-and-learn" alcoves built into spaces and a centrally-located e-café where faculty, staff, students, and industry partners can gather to share ideas.

Achieving LEED Gold brought together a number of green building elements – elements also used to educate students and visitors about sustainability and renewable energy:

» Natural ventilation provides fresh air to interior spaces
» Energy-efficient building systems require 35 percent less energy than code minimum
» Atrium daylighting supplies classrooms, labs and offices with natural light, cutting energy costs by as much as 40 percent

2. Formerly known as Portland General Electric Green Building Services

SITE PLAN

1   Reduce hardscape/shaded surfaces
2   Native, drought-tolerant plants
3   Rainwater is filtered through landscape plants to
    remove contaminants.
4   Alternative transportation: building sited near campus
    bus line, bicycle parking and showers provided.
5   Reflective roof coating used to reduce heat gain
6   Skylights
7   Photovoltaic and solar water panels

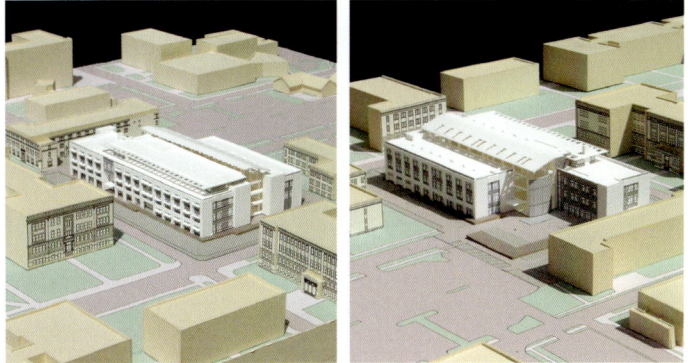

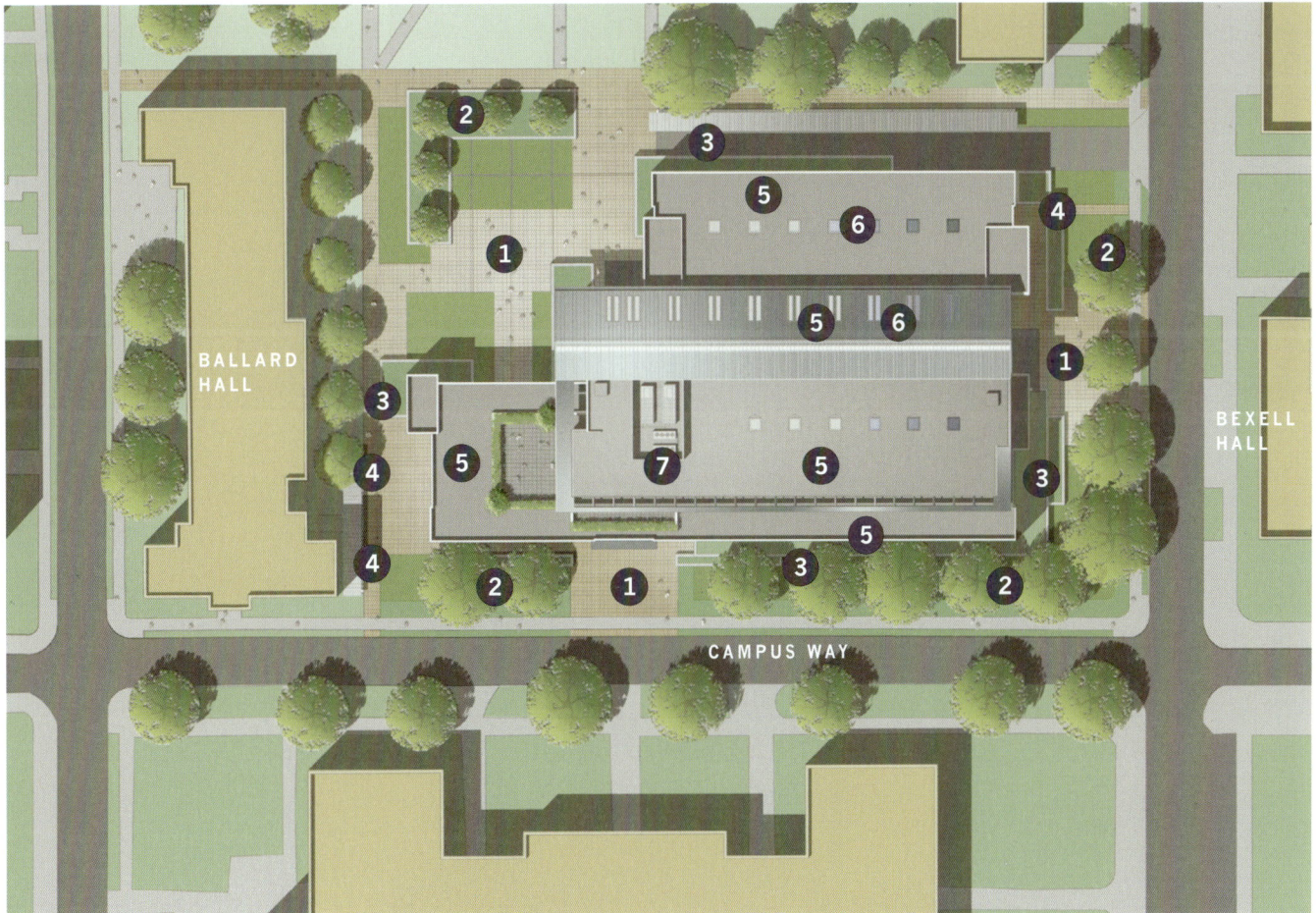

BEYOND THE BUILDING: As the first structure in a planned engineering precinct on campus, landscaping, roof materials and other measures for Kelley Engineering Center serve to minimize irrigation and offset the heat island effect.

"The Kelley Engineering Center is a catalyst for collaboration, a building designed to fuel innovation."

— *Ron Adams, OSU Dean of Engineering*

» Selected building materials feature high recycled content, much of it from local or regional sources, including the fly-ash concrete, steel, carpet, acoustical suspended ceilings, masonry, glass, and gypsum wallboard

» Zero or low VOC finishes, fiberboard and flooring located throughout the building eliminate off-gassing; these elements include formaldehyde-free wheat board, foam insulation and carpet systems, and odor-free asphalt products

» More than 90 percent of construction waste was recycled

Promoting collaboration remains the best, highest purpose of OSU's Kelley Engineering Center. With the atrium as the center point, offices, labs and other work spaces were placed according to their circulation and communication requirements. Efficiency was critical as well, to accommodate as many as 155 faculty and graduate student offices, 2,200 open computer spaces for department students, twelve conference rooms, two large theater-style classrooms, two "reconfigurable" class/conference rooms, and nine seminar classrooms.

Equally important to the building's overall design is flexibility to accommodate the changing needs of research teams, faculty and students. The floor plan represents a new approach for OSU, a break with tradition where labs are typically dedicated to individual faculty members during the course of an entire career at the university. Instead, each lab at Kelley is part of a "research-learning suite" of faculty and GRA offices assigned to a specific research project; upon completion of each project, a new team moves into that space and each area is reconfigured once again.

MEP DESIGN PROCESS

Central to its performance, the Kelley Engineering Center is organized around the atrium, with perimeter office spaces all naturally ventilated; air flows from the outside, gains heat as it passes over people, computers and other machines, then rises and exhausts out the top of the atrium. To achieve ultra-high energy efficiencies, the building also employs a combination of strategies including night flushing, heat/cold sinks, underfloor air, and ample natural lighting as well as stringent construction controls in the building shell and underfloor plenum. The project team used computational fluid dynamics computer algorithms to create the natural ventilation scheme and a daylight lab to model year-round solar exposure levels and glare. In combination, these MEP strategies resulted in first-year occupancy energy use of less than 40 percent of state energy requirements, well below the initial target of 50 percent.

Automation also plays an integral role in building operations, particularly the occupancy sensors for both climate and lighting control. In perimeter offices, for example, a sophisticated daylight/dimming system minimizes energy use by providing light only when required. In addition, occupants manage their own space heating and cooling with motorized operable windows.

Automatic controls sense when individual windows are open and adjust HVAC systems for maximum energy efficiency; likewise, a system override closes all windows in the event of bad weather and opens windows as part of the daily night flushing scheme.

While the building exceeds many key performance metrics, the University, Glumac and the rest of the design team acknowledge several lessons learned. Modeling was based on the school's original projection that it would be occupied approximately twelve hours a day, 7 a.m. to 7 p.m. Double occupancy was planned for most offices, which would have meant more people using the space, but for shorter periods of time. In reality, however, the building is occupied around the clock; graduate researchers with non-traditional schedules enter and exit the building frequently, especially late in the day and throughout the night. Due to the change in usage, the University had to reassess its projected energy use, particularly in regards to heating. Despite this increased use, Glumac still calculates the building as performing 40 percent more efficient than code.

INDOOR AIR

Per Glumac's natural ventilation design, air flows from the Center's perimeter offices (for faculty and graduate research assistants) and basement into the atrium, which acts as a stack-driven chimney, and then out through external exhaust louvers. During night flushing, all office windows and transoms open automatically to move air through offices spaces; meanwhile, atrium louvers open and push a large volume of air through the building, resulting in overall cooler temperatures.

All offices, laboratories and other work-spaces take advantage of underfloor air, which also aids with indoor air quality and provides some temperature/comfort control for occupants as well as energy savings. For Glumac, the Center served as the first in a series of successful applications of these systems in buildings. The plenum offers approximately eighteen inches of underfloor space, resulting in no overhead HVAC ducting. This scheme allows for very low pressure, 0.05 inches (.127 cm) of loss from outside the floor to the floor, compared with a medium pressure duct system in the ceiling at 1.5 inches (3.81 cm) of loss. With a goal of 80 percent pressure-holding efficiency on the floor system, this design achieved 95 percent, with only 5 percent air loss.

The building's design also includes a carbon monoxide monitoring system with operational adjustments.

WATER & WASTE

Taking advantage of Oregon's abundant rainfall, the building relies on a rainwater harvesting system for sewage conveyance and on-site irrigation, effectively reducing water use by an estimated 372,000 gallons (1,408,173 liters) per year. In winter, rain is channeled down from the Center's roof through the building's landscaping planters into a 15,000-gallon (56,780-liter) greywater reclamation system for use in toilet flushing and urinals.

ALL-NIGHTER: Building performance relies heavily on night flushing operations to move a large volume of air through office spaces and the atrium, aiding in indoor air quality and cooler temperatures.

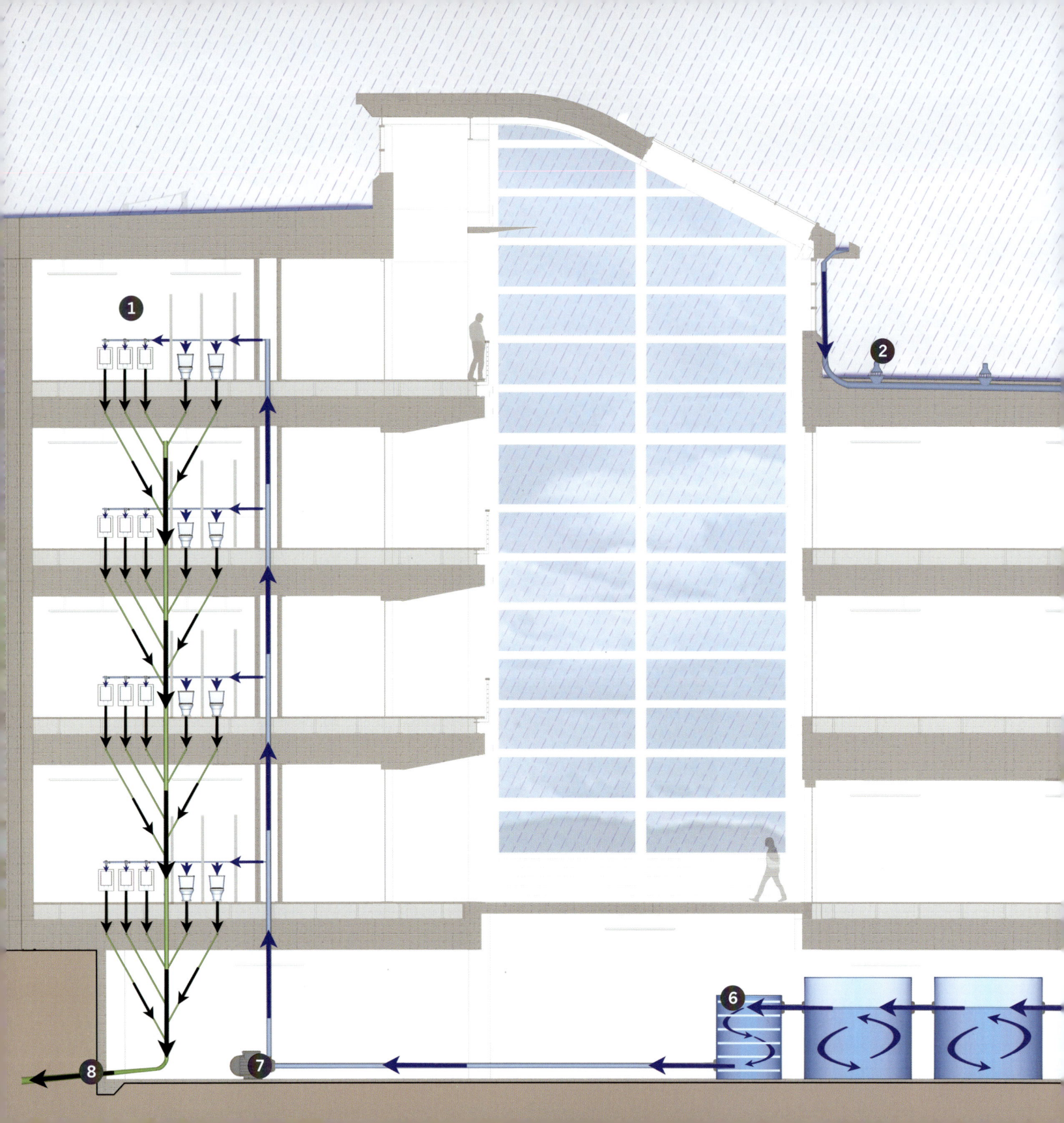

RAINWATER RECLAMATION

Harvesting rainfall on site yields an estimated 372,000 gallons (1,408,173 liters) for sewage conveyance and irrigation – with all water channeled from the Center's roof through the building's landscaping planters into a 15,000-gallon (56,781 liter) greywater reclamation system.

1   Toilets and urinals flushed with harvested water
2   Rooftop water collection with roof drains
3   Downspout
4   Rainwater discharge into vegetative filter
5   Filtered water into collection: three 5,000-gallon (18,930-liter) tanks
6   Ultraviolet filtration system
7   Pumps supply harvested water to toilets and urinals
8   Black water discharged to sewer

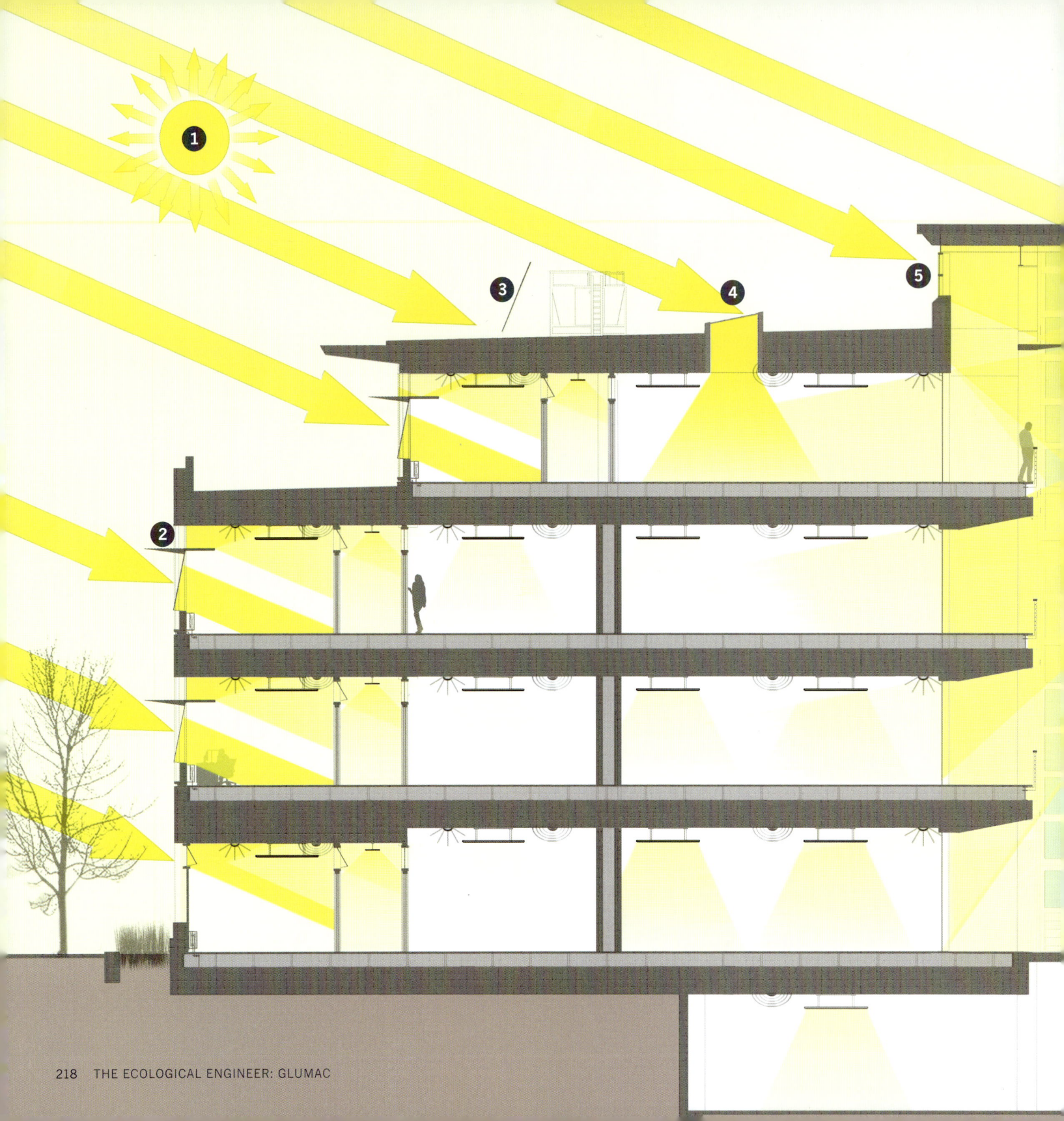

DAYLIGHTING

Natural light floods the atrium with 25 foot-candles (270 luxes) even on overcast winter days. Within classrooms, laboratories, offices and throughout the rest of the Center, photo and occupancy sensors modulate light levels based on the amount of daylight present as well as the movement of occupants.

1 Direct south sun
2 Light shelf and north sun shade
3 Photovoltaic array / mechanical screen
4 South skylight
5 South clerestory window
6 Atrium skylight
7 North skylight
8 Occupancy sensors
9 Photo sensors
10 Indirect north sky dome light

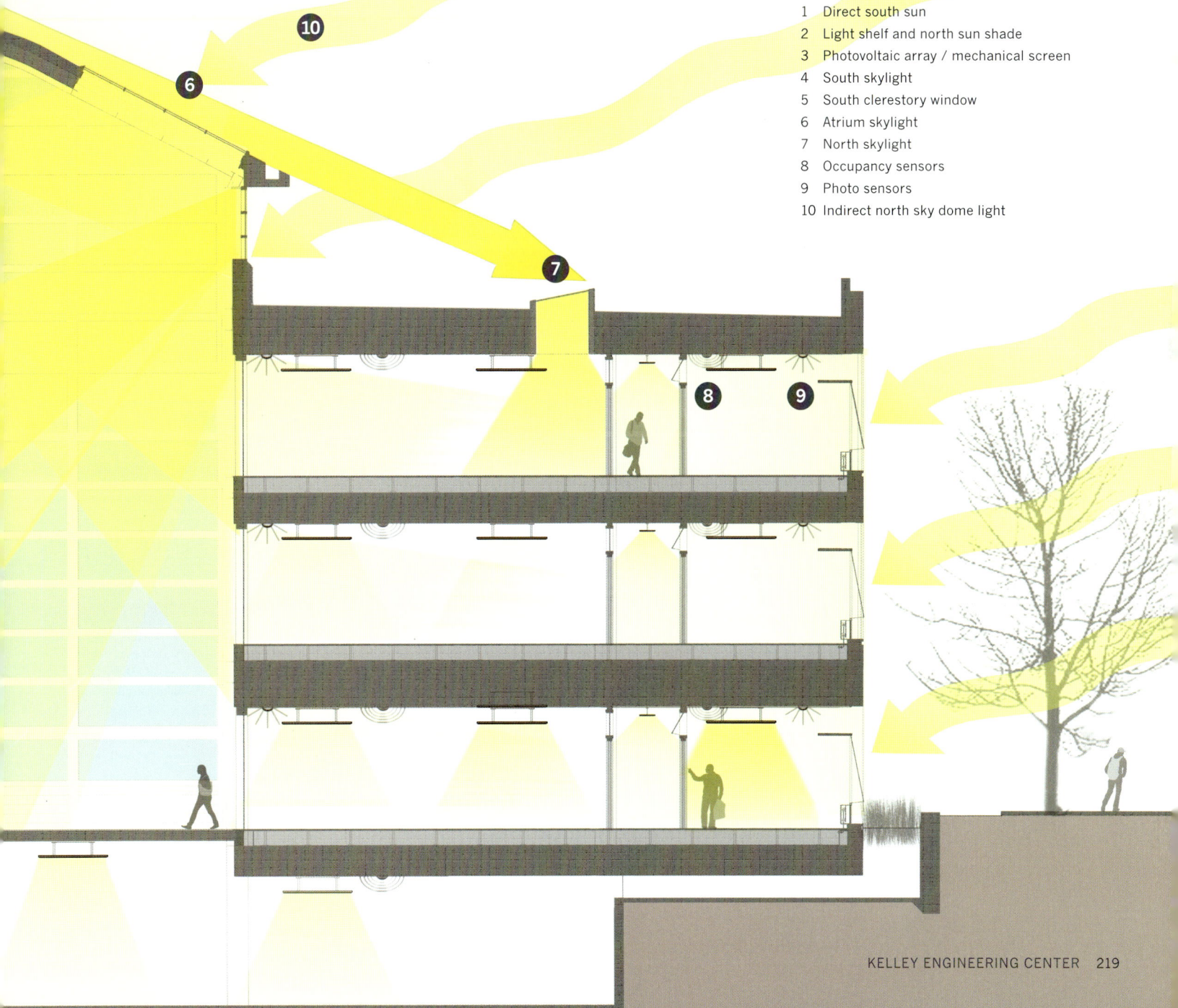

OFFICE HOURS: In faculty offices and graduate research assistant (GRA) work areas along the building perimeter, occupants can manage their own comfort through a series of lighting, natural ventilation, heating and cooling measures.

1 South side exterior sun shade
2 Light shelf
3 Fresh air intake grille and duct
4 Computer-controlled motorized windows & transoms
5 Environmental control system operation panels
   (HVAC, windows, lights)
6 Manually adjustable floor diffuser

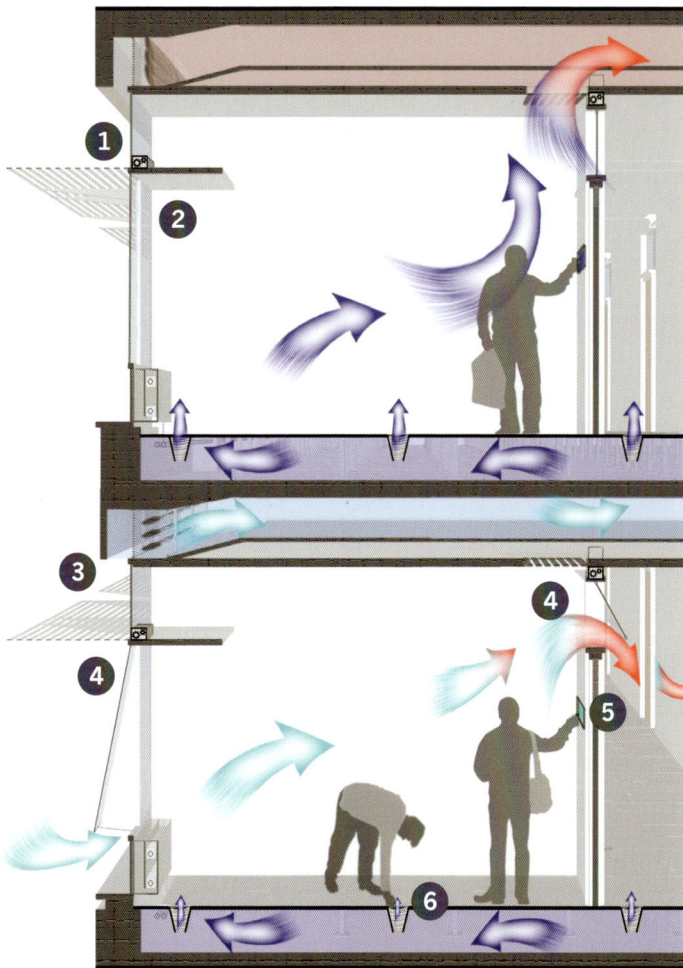

Water-efficient fixtures also reduce the building's water usage. Combined with rainwater reclamation, the building reduces the baseline potable water usage by 65 percent.

HEATING & COOLING

Glumac's design employs multiple zones of natural ventilation to heat or cool the Kelley Engineering Center 80 percent of the time; in particular, the four-story atrium maintains its own temperature, exhausting air in warmer weather and drawing in cool air to minimize air conditioning as necessary. Except for laboratories and server rooms, the building requires mechanical heating (campus steam) and cooling (chillers) only 20 percent of the time.

Interior offices and laboratories, which contain a large number of computers and peripheral equipment, receive more intensive use, even late at night. To compensate, these spaces feature variable speed diffusers built into the raised access flooring. The underfloor plenum delivers a constant temperature, using semi-displacement to mix the air and gradually move it into a comfort range of the first four or five feet (1.2 or 1.5 meters) rather than attempting to mix an area's entire volume of air. As a result, this system provides cooling or heating with a five-degree (2.8° C) differential in temperature versus fifteen degrees (8.3° C) required from overhead ductwork. The raised access floors also allow for easy maintenance and upgrades to electrical cabling.

"For a while, this building wasn't widely accepted and was referred to as kind of an 'opulent building and too technically advanced.' People questioned why we were spending money on collecting rainwater in the Willamette Valley. Now the atrium is probably one of the most spectacular spaces at OSU – people speak of Kelley as 'Wow, why can't we do something like that again?'"

— *John Gremmels, Senior Project Manager. OSU*

In addition, occupants may heat or cool their offices by pressing a button to open or close a window, which then disables mechanical heating and cooling in the space. In cooler weather, building controls respond to an open window by turning off the heat supplied to that room. All rooms include a fin tube radiator behind the operable windows, also tied to the automated system. Finally, the building's Energy Star-rated, high albedo roof deflects radiated heat overhead.

LIGHTING & DAYLIGHTING

The Oregon sky supplies plenty of natural light into the building, providing 25 foot-candles (270 lux) within the atrium even on overcast winter days. At each end, mosaics of translucent glass panels reduce glare, constantly changing in appearance. Daylight also illuminates virtually all classrooms, laboratories and offices adjacent to the atrium and windowed exterior walls, cutting electricity consumption for lighting up to 40 percent. Lighting design also relies on south-facing exterior sunscreens to minimize heat gain along with interior light shelves to control glare.

To augment daylighting in offices and other interior spaces, designers specified indirect linear light fixtures to meet requirements for work or general illumination. A sophisticated daylight/dimming system also controls lighting energy use by providing lighting only when required. The atrium itself contains no lighting fixtures other than floor lamps which appear as street lights.

SITE

The Center's site design reduces storm runoff, so that uncollected rainwater is filtered through native plantings to remove contaminants. The outside plaza also features permeable tiling to reduce the amount of water flowing into city storm drains.

ENERGY

Overhead, the building roof features a 2.4 kW photovoltaic array that doubles as the cooling tower screen wall. This solar panel, intended primarily as a demonstration unit and teaching tool, supplies a small amount of power back into the utility grid as part of the school's net metering plan to reduce electrical energy costs. An evacuated tube solar hot water collector system also provides domestic hot water for the building at 70 percent utilization efficiency.

In addition to photovoltaic generation, the university purchases a portion of the building's electricity from renewable wind, solar and biomass sources.

COMMISSIONING

Following construction, a third-party conducted thorough building commissioning to ensure that mechanical and electrical systems would work as designed. Glumac provided further system review and building operator training to ensure efficient ongoing operation.

HEATING, AIR CONDITIONING AND VENTILATION

TOTAL BUILDING HVAC: Powered by its 74-foot-high, glass-paned atrium, Kelley's passive design scheme uses multiple zones of natural ventilation to heat or cool the building 80 percent of the time, with mechanical heating and cooling systems supplying the remainder.

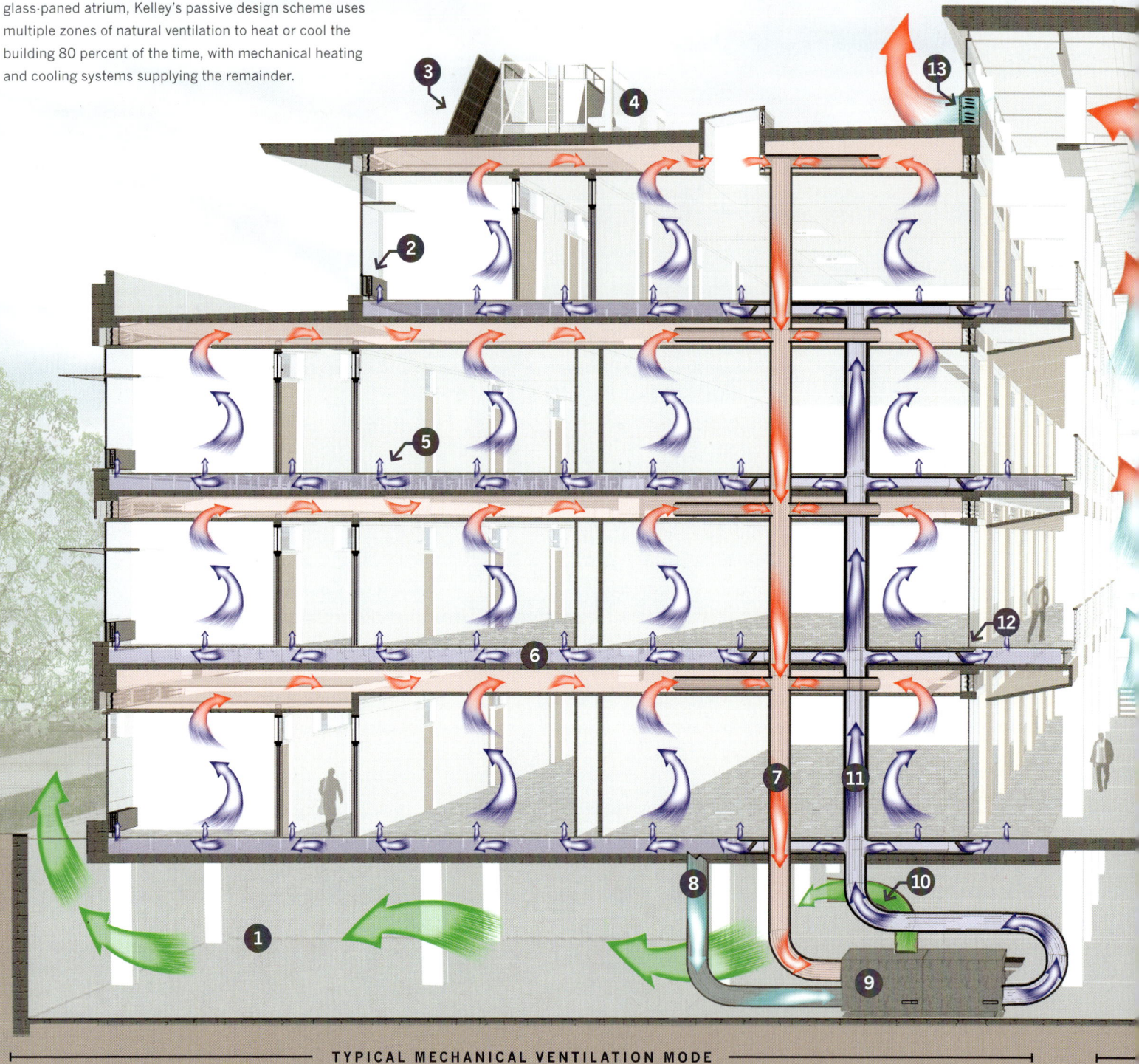

TYPICAL MECHANICAL VENTILATION MODE

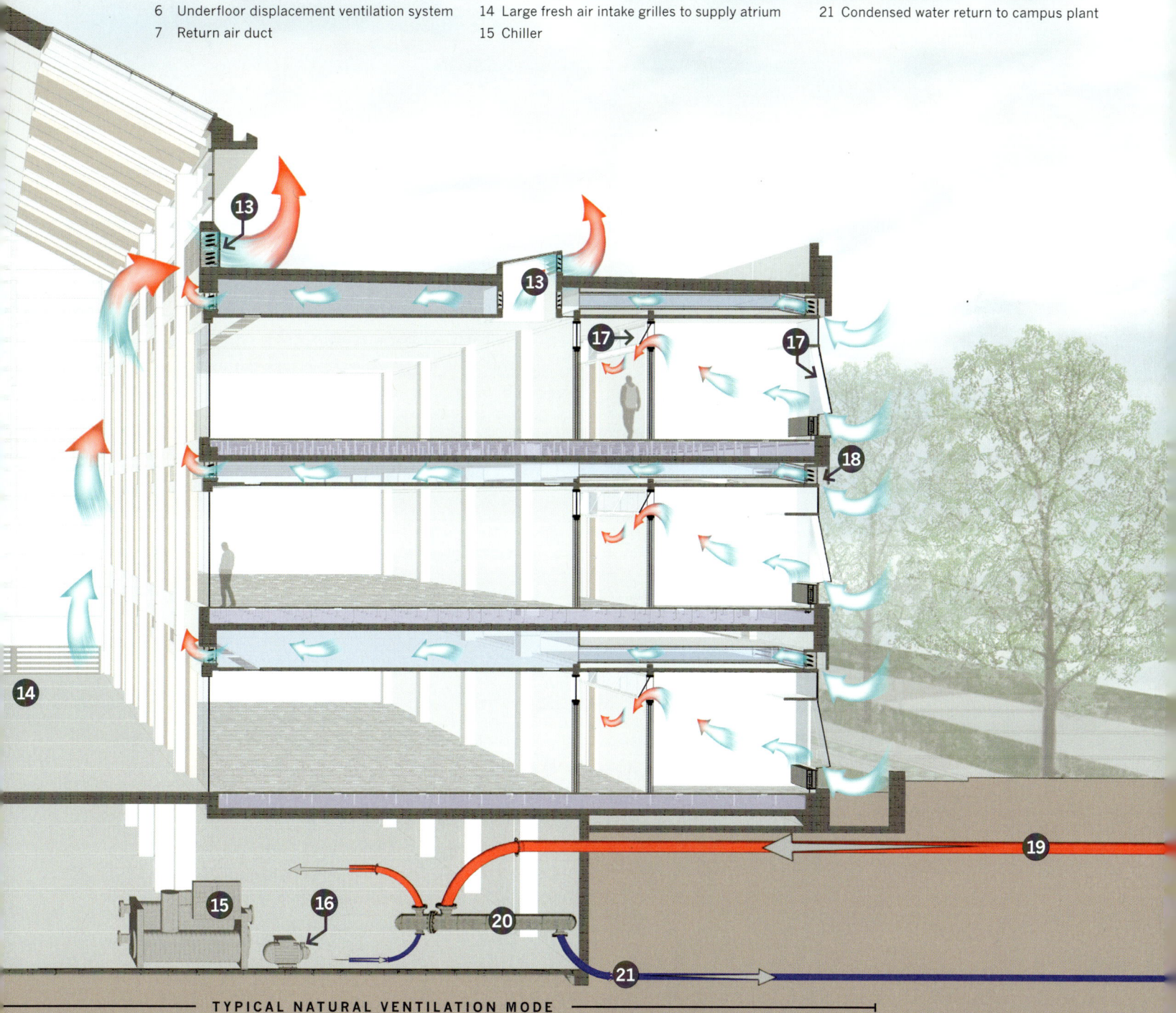

1   Basement exhaust plenum relief to outside through ground grille
2   Perimeter hydronic baseboard heaters
3   Photovoltaic array/mechanical screen
4   Cooling towers
5   Individually adjustable floor diffusers
6   Underfloor displacement ventilation system
7   Return air duct
8   Fresh air supply intake shaft from roof
9   Air handler
10 Relief air out into basement exhaust plenum
11 Supply air duct
12 Variable air volume (VAV) boxes
13 Exhaust grille for natural ventilation relief
14 Large fresh air intake grilles to supply atrium
15 Chiller
16 Chilled water/heating water pump
17 Manually operable/automated motorized windows and transoms
18 Fresh air intake grille and transfer duct
19 Supply steam from campus plant
20 Steam/hot water heat exchanger
21 Condensed water return to campus plant

TYPICAL NATURAL VENTILATION MODE

BUILDING PERFORMANCE

Within a year of its opening, OSU facilities staff began to notice a widening gap between energy projections for the building and actual performance. Increasingly one of the most popular locations on the north side of campus, Kelley Engineering Center stayed open longer while attracting more students than anticipated. "Originally, we programmed the building to operate twelve hours a day, for staff, undergraduates and graduate students to take their classes and leave," explained Steven Straus. "But it's such a nice space that people really want to be there – and the place is energized day and night."

As a result, OSU asked Glumac to investigate the underlying reasons for significantly higher energy consumption over that projected by the State Energy Efficiency in Design (SEED) energy study prepared during the building design. Glumac's energy modelers focused on two related areas: 1) review of actual building operations for possible sources of higher energy use and 2) energy modeling reconciliation, to review/compare actual operating conditions against input assumptions from the original SEED energy model and revise energy models accordingly.

First, Glumac plotted total energy and steam consumption for the SEED model versus actual use between May 2006 and October 2007. This analysis uncovered its biggest deviation: total steam use estimated at 1,088,300 pounds (493,600 kilograms), compared to the actual utility bill at 6,131,700 pounds (2,781,300 kilograms) of steam during this period. Steam generation within the building covers both heating hot water and domestic hot water. The revised modeling, however, concentrated primarily on the much larger energy load for heating hot water in three areas: at central air handler hot water heating coils, at VAV system reheat coils, and at office radiators.

Model alignment and retro-commissioning efforts then examined these sources and the potential impact of subsidiary systems affecting energy use. Specific changes to the model included:

» Building servers/computer room load: using 25 percent and 33 percent of the original modeled electric load to reflect actual conditions

» Private office/lab space plug loads: using 25 percent of the original model, since these rooms contain just 1 CPU as opposed to the original 4 CPUs planned per room

» Lighting: revised to increase the operating hours based on observations

» UFAD system control: floor diffusers opened up 100 percent of the time, as opposed to a varied schedule allowing diffusers to be open between 40 and 100 percent of the time

» Fume hoods: added two 3,000 cfm units (1,400L/s) to operate continuously

» Air handler operating schedules: adjusted fan schedules, increased limits on reheat temperatures for several VAV

systems to 90°F (32°C) with warm-up cycle discharge temperatures at 95°F (35°C); modified morning warm-up cycle to simulate actual found warm-up conditions matching the approximate loads of 3,000 pounds (1,400 kilograms) per hour during the coldest days of the year, and eliminated heat recovery chiller operation

The above changes also aligned the energy model to within 5 percent of actual energy consumption. Actual building use and reconciled model energy are now 17 percent better than facilities found within the Center for the Built Environment's database for colleges and universities. Savings between a revised SEED baseline and the as-designed model remain relatively the same as in the original SEED study – essentially a total energy reduction of 30 to 35 percent compared to the code baseline building. To further improve building performance, Glumac designers identified several strategies for reducing steam consumption in space and ventilation heating. In addition, they recommended retro-commissioning of building operations to include verification of natural ventilation and outside air damper control sequences.

ALIGNING THE ENERGY MODEL

Once OSU discovered that students were using the new building 24 hours a day – rather than 7 a.m. to 7 p.m. as initially projected – Glumac worked to reconcile the model with actual energy consumption and improve operations through retro-commissioning efforts.

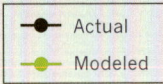

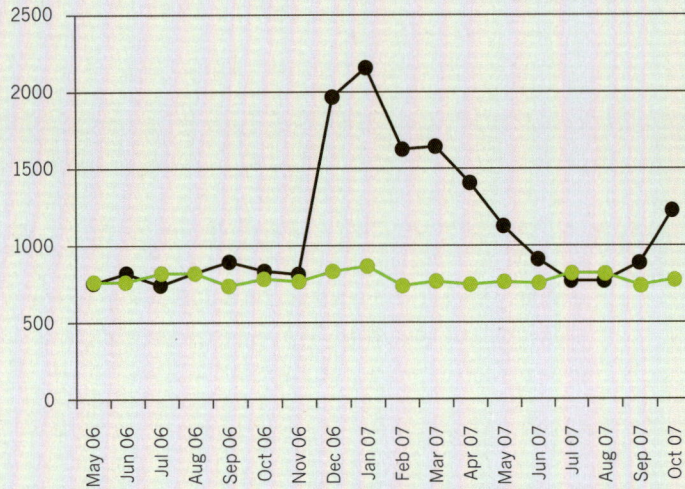

Legend:
- ● Actual
- ● Modeled

TOTAL ENERGY USE:
Original model vs. Actual consumption

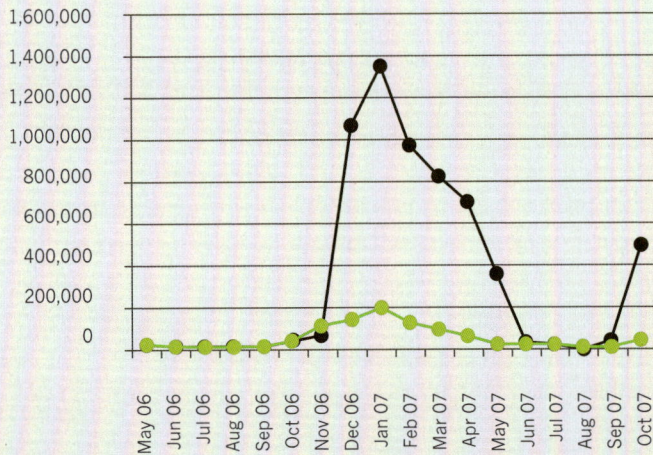

STEAM CONSUMPTION:
Original model vs. Actual consumption

TOTAL ENERGY USE:
Revised model vs. Actual consumption

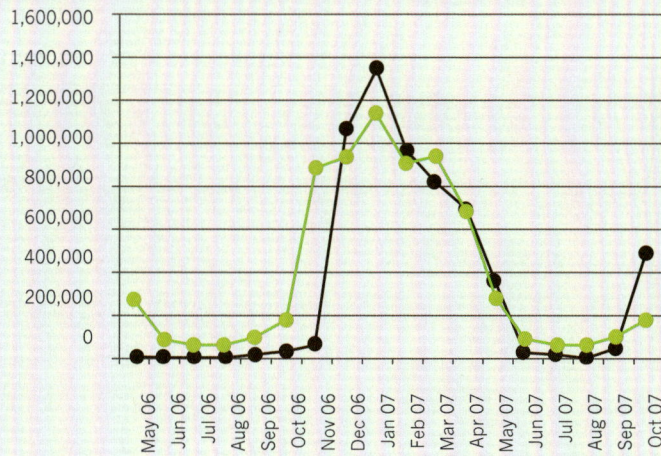

STEAM CONSUMPTION:
Revised model vs. Actual consumption

BREWERY BLOCKS

PORTLAND, OREGON

Within the Pearl District, one of the nation's finest examples of urban renewal, the Brewery Blocks serve as the anchor – a landmark achievement in mixed-used development and sustainable design. Centered around the historic Blitz-Weinhard Brewery (built in 1906), this five-block neighborhood forms a gateway to downtown as home to urban retail, creative Class A office space and residential housing. Block One houses a rooftop high-efficiency chilled water plant that supplies cooling to all Brewery Blocks buildings. Block Two, surrounding the Brewhouse and Tower, features chilled air HVAC, natural ventilation and daylight harvesting. Block Three, "The Henry," a 15-story high-rise adjacent to the historic Gerding Theater, incorporates heat reclaim, optimized glazing and sun shades. Located on Block Four, the 294,000-square foot (27,300-square meter) M Financial Plaza contains a chilled air HVAC system, daylight harvesting and building-integrated photovoltaic panels, all combining to reduce energy use to 30 percent below ASHRAE standard 90.1.

"...the Brewery Blocks are quintessential Portland. They fit into their context, simultaneously evoking the historic character of the neighborhood while also striving for a more contemporary identity."

— *Brian Libby*
ArchitectureWeek, *May 11, 2005,*
from "Mixed Use Brewery Blocks",
Page E1.2

PROJECT DETAILS

5 buildings, 1.7-million-square feet (158,000-square meters) of mixed-use office, retail and residential buildings, underground parking

Completed 2003

Architect: GBD Architects

Contractor: Hoffman Construction Company (Blocks 2 & 4), R & H Construction (Block 1)

Developer: Gerding Edlen

"The most important element of the Brewery Blocks project was the integrated design process itself – used on a broader scale than Glumac or any others had done before. We had a role in every element of the development, from peer review and consulting to full design. And, as a result of its success, other developer-led projects have followed that model."

— *James Thomas, Vice President, Glumac*

THE BLOCKS

Block One: Whole Foods Building, LEED Silver certified

Block Two: Brewery Blocks Brewhouse & Tower, LEED Gold certified

Block Three: The Henry Building, LEED Gold certified

Block Four: M Financial Plaza, LEED Gold certified

Block Five: The Louisa Building, LEED Gold certified

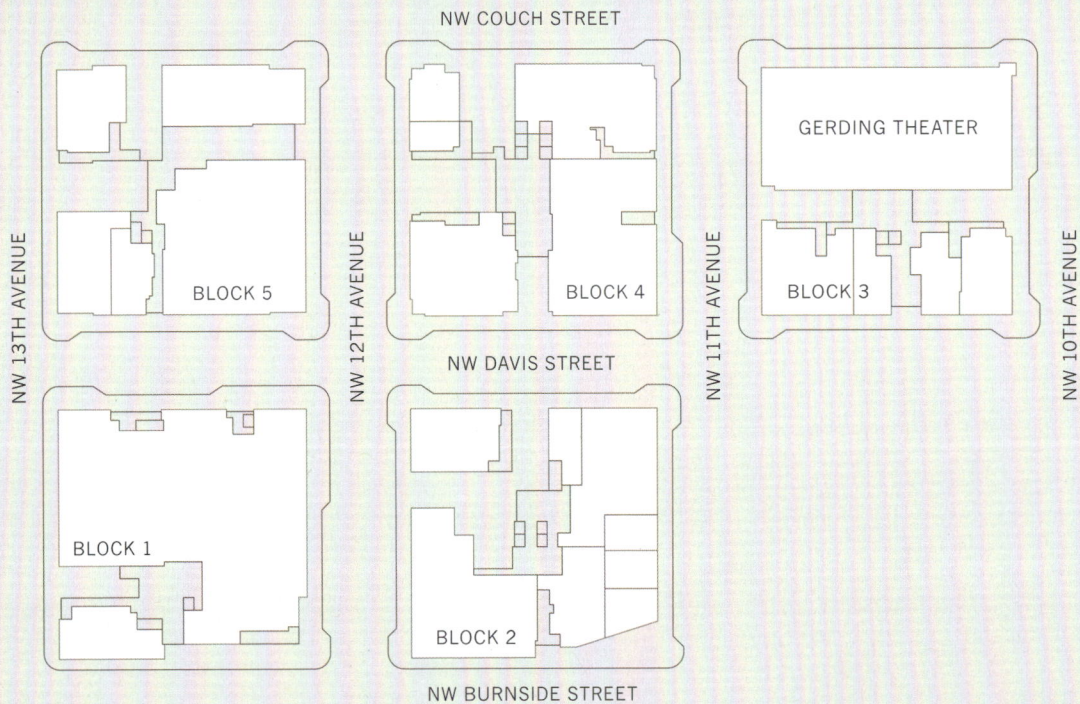

NW COUCH STREET

GERDING THEATER

BLOCK 5

BLOCK 4

BLOCK 3

NW 13TH AVENUE

NW 12TH AVENUE

NW 11TH AVENUE

NW 10TH AVENUE

NW DAVIS STREET

BLOCK 1

BLOCK 2

NW BURNSIDE STREET

"For Portlanders, the Mercy Corps headquarters will be a destination, a source of pride and a model of environmental sustainability. For Mercy Corps, it is the business opportunity of a lifetime."

— Mercy Corps, "A Reflection of our Values"

MERCY CORPS HEADQUARTERS

Teeming with life – an apt description for the activity and passion inside Portland-based Mercy Corps, an international relief-and-development agency working in thirty-eight nations across the globe. Supporting that mission became a central goal in creating the non-profit's new headquarters, in part by consolidating its staff and operations into a single office and offering a unique venue for public engagement. From top (green roof, structured for future PV) to bottom (parking lot with pervious pavers), the project highlights sustainability as well, with recycled materials and period-appropriate, energy-efficient windows throughout. Additional design features include a full-building heat pump system, daylighting controls, a roof monitor with motorized openings to induce natural ventilation, a dedicated outside air system with CO_2 sensors, and a switchable glass façade on the south side to shade the atrium and Action Center.

"Because they are there to serve people – and their vision for change is to secure productive and just communities – the Skidmore Fountain Building became the perfect place for Mercy Corps. Highlighting their core philosophy is the Action Center on the first floor, a space where people can be introduced to the organization and even video conference with families in Africa or Southeast Asia who need their support. Every system is at work here: daylighting, natural ventilation, comfort – really, it is the core of Mercy Corps."

— *Rob Schnare, Glumac*

PROJECT DETAILS

Four-story, 80,000-square foot (1,400-square meter) office building

Completed June 2009

Architect: THA Architecture

Contractor: Walsh Construction Company

Owner / Developer: Mercy Corps International

LEED Platinum certified

WAYNE L. MORSE
COURTHOUSE, 2006: Inside
and out, the streamlined
design of the federal
courthouse speaks to honor,
tradition, safety and as a place
for people – all within a setting
that looks to the future.

CHAPTER SEVEN:

WAYNE L. MORSE UNITED STATES COURTHOUSE

AT THE HEART OF JUSTICE

"A courthouse should allow transparency for a seemingly obscure process. The building should be transparent. The Eugene courthouse is designed to allow the beauty of northwest light to fill the interior and give every citizen a view to the inside and with it ownership of the judicial process."

— *Judge Michael Hogan*
U.S. District Court for the District of Oregon, Wayne Lyman Morse United States Courthouse, U.S. General Services Administration, December 2006

"The great and invigorating influences in American life have been the unorthodox: the people who challenge an existing institution or way of life, or say and do things that make people think."

— William O. Douglas
Points of Rebellion, New York: Random House, 1969

EUGENE, OREGON

The building site is located along the city's northeast greenway not far from pedestrian paths and sprawling bike trails. The building's sweeping metal and glass façade overlooks the Willamette River, a commanding presence beside a busy highway viaduct at the edge of central Eugene. Outside are breathtaking views of the Cascades. Inside its three curving pavilions, the courtrooms and judges' chambers link through the two-story atrium to a complex of offices. The Wayne L. Morse United States Courthouse expresses transparency, access and civic engagement in its architecture while also being reminiscent of the old landmarks once found on every town square.

Completed in 2006, the $78 million, 270,000-square foot (25,084-square meters) building displays five stories above grade with one below grade. The first and second floors hold offices for the courts and their clerks, the U.S. attorney, probation and pretrial services, the U.S. Marshals Service, the U.S. General Services Administration (GSA), two U.S. senators, and one member of the U.S. House of Representatives. The courthouse also serves as the nucleus of a small district of mixed-use warehouse renovations, within a half-mile of three high-density residential neighborhoods close to rail and bus lines.

As part of the Ninth Judicial Circuit's District of Oregon, the new courthouse and its predecessor have witnessed some of the nation's most significant environmental federal rulings. These rulings include protecting the northern spotted owl by blocking new timber sales in old growth forests in the Northwest, upholding the right of citizens to sue under the Clean Water Act, and voiding hundreds of grazing leases that threaten Southwestern national forests.

"When U.S. District Judge Michael Hogan first saw designs for the new federal courthouse in Eugene, OR, he hated them. The fantastically modern building, with its undulating glass and stainless-steel façade, simply did not conform to his notion of what a courthouse should look like. ... A disappointed Hogan immediately challenged the architect on numerous points. To his surprise, Mayne challenged him right back."

— *Jay W. Schneider*
"Courthouse Pushes the Boundaries of Tradition: Wayne L. Morse U.S. Courthouse," Building Design and Construction, *May 1, 2008*

Reconciling security and sustainability posed the biggest design challenge for the project. The GSA sought LEED Silver certification at a minimum to showcase the latest in sustainable architecture, engineering and construction. Security, however, was the top priority. As a Level IV facility – just one level below buildings such as the Pentagon and CIA Headquarters – courthouse tenants would include high-risk law enforcement and intelligence agencies in addition to the courts and judicial offices that house highly sensitive government records.

Both goals were achieved and largely exceeded through a whole-building, integrated design approach led by Glumac, the Portland office of JE Dunn Construction, DLR Group, and Morphosis as the design architect. Delivered as the nation's first new LEED Gold Federal courthouse, the building is 38 percent more energy efficient than conventional construction – highlighted by innovative strategies that include fan-wall ventilation, underfloor air, radiant-floor heating and cooling, and advanced building automation. Together with optimum building security, daylighting, fresh air, and organic forms throughout the courthouse, its carefully-crafted judicial spaces succeed at a healthy, human scale.

Named in honor of Wayne L. Morse (1900-1974), a U.S. Senator and Dean at the University of Oregon School of Law, the building also became the first within GSA Northwest Arctic Region 10 to achieve LEED certification. And just a year after its opening, the AIA Committee on the Environment declared the courthouse a Top Ten Green Project for 2007.

ENVIRONMENTAL OBJECTIVES

Although courthouses are traditionally located within a city center, the GSA pursued a completely new direction for the building. City officials wanted to forge a stronger connection between downtown Eugene and the river, so they offered the site of an abandoned cannery in a small waterfront warehouse district. A four-acre (1.6-hectare) brownfield, the project's site was covered with metal-roofed structures and asphalt, all contributing to an urban heat-island effect and extensive storm water runoff. With annual rainfall averaging fifty-one inches (129.54 cm), restoring percolation rates to the site became another key goal for the project.

The city also felt the new building could help drive redevelopment, with numerous plans in the works to turn the district into a high-density, mixed-use neighborhood. Existing public rail and bus lines now connect the courthouse to the larger community, while the project team worked with the city to create new public-transit stops nearby. Preferred parking is available for employees and tenants using low-emitting and fuel-efficient vehicles.

DESIGN OVERVIEW

As conceived by Thom Mayne of Morphosis, the design wraps the five-story courthouse in ribbons of 16-gauge (1.3-millimeter) stainless steel, its upper floors divided into three curving pavilions with a pair of courtrooms (two district courts, two magistrate courts, and two bankruptcy courts) in each pavilion. The floors below serve as a plinth for the more distinctive architecture above, housing administrative and support spaces, offices for the courts, the U.S. Attorney, probation and pretrial services, the GSA, and elected officials.

Courthouse architecture itself often becomes the subject of intense debate. Iconography and the need for a tangible expression of values and beliefs were very much on the mind of Michael Hogan, then chief judge of the U.S. District Court when a new federal building for Eugene was first considered. In 1999, Hogan lent his opinion to narrow the list of finalists for the courthouse design competition; ultimately the GSA, the building's owner, selected Mayne and his Santa Monica-based design firm.

Hogan wanted a traditional design where Mayne focused on ideas rather than shapes. Creative tensions between the judge and the architect over many months turned into a unique dialogue and then collaboration on blending classical architecture with modern materials and technologies.

Thom Mayne's sketches began with the courtrooms, which range in size from 1,800- to 3,000-square feet (170- to 280-square meters). His design resulted in curving, organic forms that taper toward the front, while opening sightlines to focus to the judges' benches. Cherry and walnut paneling wrap around the upper portion of the rooms. He also relocated the jury assembly room to the second floor, where it now doubles as a multipurpose space for art exhibits and community meetings.

First and foremost, Mayne wanted to make the building's security transparent, creating an inviting approach for users and visitors. Simultaneously, the design still had to meet stringent security and site perimeter requirements to protect against bombings as well as ballistic, biological and chemical attacks. Grassy terraces, planted berms and adjacent amphitheater seating all function as security barriers. Periodic wall openings on the south lead to an internal public plaza at ground level and up a set of massive stairs to the main entrance on the second floor. Raising the entry level – along with locating courtroom pavilions set back from surrounding streets and the secure underground parking for judges – achieved additional security factors.

Through a series of green design charrettes, the project team explored a range of options for combining security, sustainability and adaptability to respond to future tenant needs. Solutions to pre-design goals quickly emerged during these sessions:

"The collaborative delivery approach no doubt contributed to the success of the project — very disciplined, particularly when dealing with the unusual shapes and forms that Thom Mayne invented. The only way we could possibly have pulled this off was to bring in the contractor during the design process. So the mechanical and electrical contractors could actually work with the Glumac designers to create a building that was within budget, technologically advanced and ensured that everybody was on the right track."

— *Richard Broderick, U.S. General Services Administration*

MOMENTOUS DECISION: Home to the Ninth Judicial Circuit's District of Oregon, the 270,000-square foot building was the nation's first new LEED Gold Federal courthouse – designed to be 38 percent more energy efficient compared to conventional construction.

» To create adaptable tenant space, the team opted for a raised access floor and underfloor air, allowing for easy reconfiguration of cables and office layouts

» An interior courtyard encourages daylighting

» Public gathering areas invite community interaction

» Underground parking preserves open space and enhances view corridors to the surrounding landscape

» Construction focused on recycled content, particularly steel and aluminum components, along with many materials selected for their regional availability, minimal maintenance needs, and low chemical emissions

Collaboration as a team was exceptional and continued throughout the design process. Designers, the owner, tenants, artisans and contractors spoke daily and participated in weekly team meetings to review project goals. By utilizing a "construction manager at risk" delivery method, the general contractor was able to participate in the design phase and provide critical input on the project's constructability, schedule and budget.

MEP DESIGN PROCESS

As with Mayne's use of form to express openness and Judge Hogan's appeal for blending sustainable with traditional elements, the MEP systems reflect a conscious effort by designers to rethink the use of judicial spaces throughout the building. It was important to apply technology that would increase visual and thermal comfort but also integrate seamlessly with the architecture. To accomplish these attributes, offices feature underfloor air and extensive use of low-e glazing. High ceilings in lobbies, corridors and other public areas are coupled with radiant floor systems. A courtyard allows natural light to infiltrate building interiors. This bioclimatic design effectively connects the courthouse, inside and out, to the surrounding landscape and community without compromising high security needs. The client's desire for a healthy, comfortable indoor environment also influenced many decisions about building automation, lighting, HVAC design and materials.

The level of cooperation among MEP team members, too, was exemplary. Early involvement by subcontractors and the general contractor led to putting cost controls in place early as well as pre-purchase of major pieces of equipment. Final documents were then designed around these early selections – a step that later proved crucial to the project's success given the limited space allotted for mechanical systems. Over fifteen months of pre-construction meetings, the project team discovered $665,000 in HVAC savings through value engineering – savings then applied to mechanical upgrades and other enhancements for the building.

ARCHITECTURAL INTEGRATION

Radiant slabs provide heating and cooling in public spaces (corridors, lobby, atrium), along with ventilation air through a baseboard wall grille hidden in a relief detail at the base of the canted wall. Underfloor air supplies heating, cooling and fresh air to all other spaces. Radiant slabs rest on a thick layer of insulation to prevent heat loss to structural concrete, while keeping the system level with adjacent raised access flooring.

1 Structural concrete
2 Rigid insulation
3 Radiant slab with embedded PEX Tubing
4 Topping (finished flooring) slab
5 Fresh air supply duct
6 Damper

7 Vertical fresh air plenum
8 Perforated diffuser grille
9 Raised access floor
10 Manually adjustable floor diffusers
11 Return air plenum in suspended ceiling
12 Return air duct

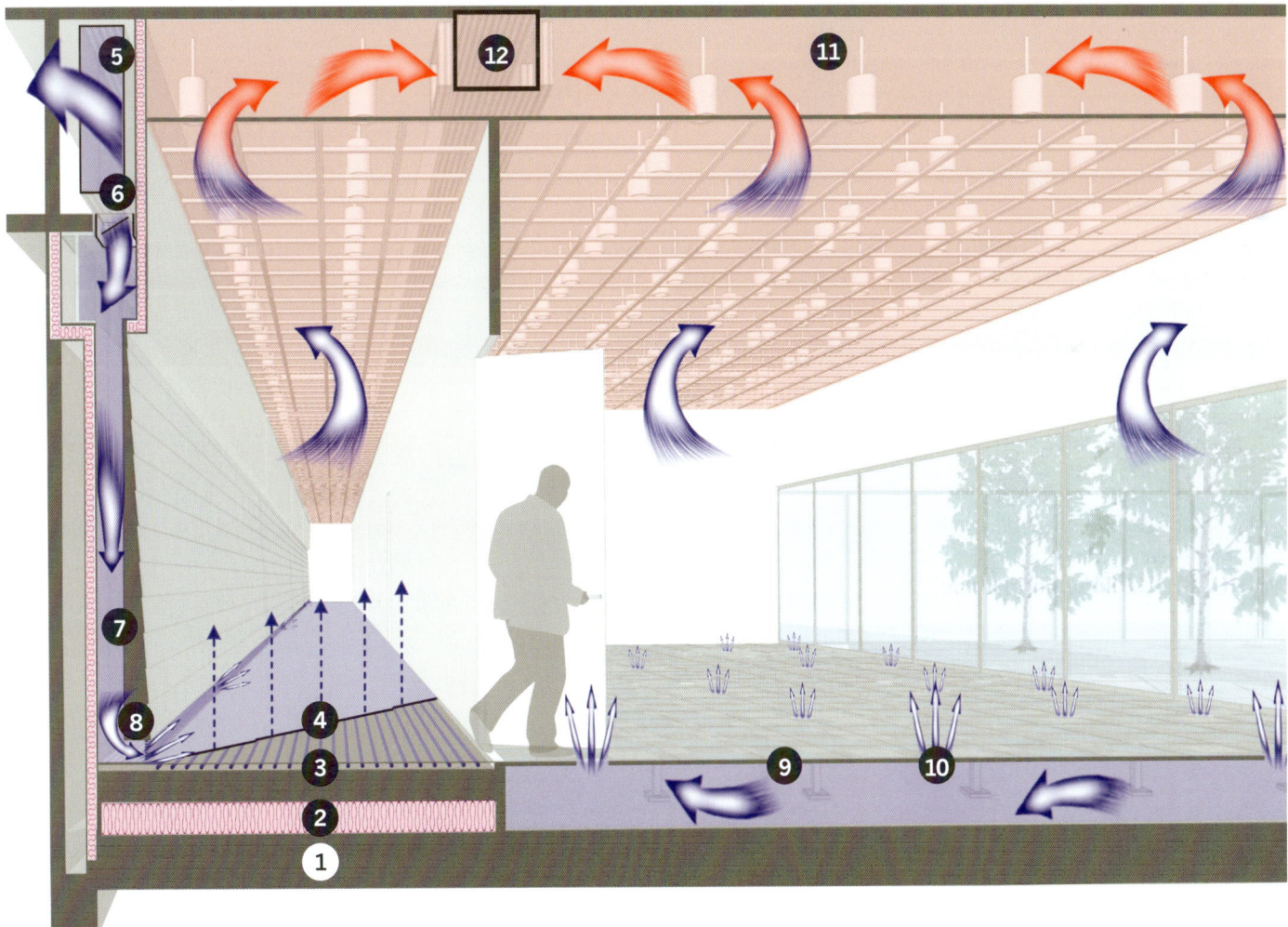

Extensive use of BIM also aided team coordination and decision making. All designers used Autodesk Building Systems to generate three-dimensional sections and isometric views of MEP systems, creating a virtual, functioning 3-D model to identify conflicts within tight building spaces before installing any equipment. Subcontractor meetings became collaborative sessions, a place to modify shop drawings in real time and carry the changes back to work crews for implementation. Through this process, designers were able to reduce two of the floors by about twelve inches (30.48 cm) without compromising overall ceiling heights. Contractors estimated that 3-D shop drawings helped to identify 95 percent of potential conflicts, versus a typical 80 percent discovery rate using standard 2-D processes.

Without question, courtroom daylighting is the highlight of this complex and demanding program – a smart design move that introduces natural light from adjoining circulation areas while addressing security requirements. Conventional design practices surround a courtroom with judges' chambers and jury rooms to create isolation for security and make use of the higher ceilings as required by court proceedings. However, by locating these rooms to the floors above, this design allowed for placement of perimeter windows along the courtroom outer walls as well as feature windows at the judge's bench, effectively saturating these spaces in daylight.

1   Return air flows into the unit from the return duct
2   Some or all the return air is exhausted from the system, depending on economizer operation
3   Fresh outside air supply
4   Warm return air mixes with fresh outside air
5   Damper controls flow of mixed air and fresh air into the supply side of the system
6   Filter bank filters all air as it passes through the system
7   Heating and cooling coils are used to adjust the temperature of the air depending on economizer operation
8   Fan walls
9   Conditioned air is pushed out into the building through supply ducts
10 Hot water supply
11 Hot water return
12 Chilled water supply
13 Chilled water return

FAN WALL AIR HANDLER SYSTEM

As a smaller and acoustically superior type of air handler, fan wall technology offers an efficient alternative to large house centrifugal fans. Common in many Glumac designs, the small fans spin slower, lowering vibration and noise production (side effects of traditional large single fan units) – with the added benefit of redundancy in the event that one fan within the system malfunctions.

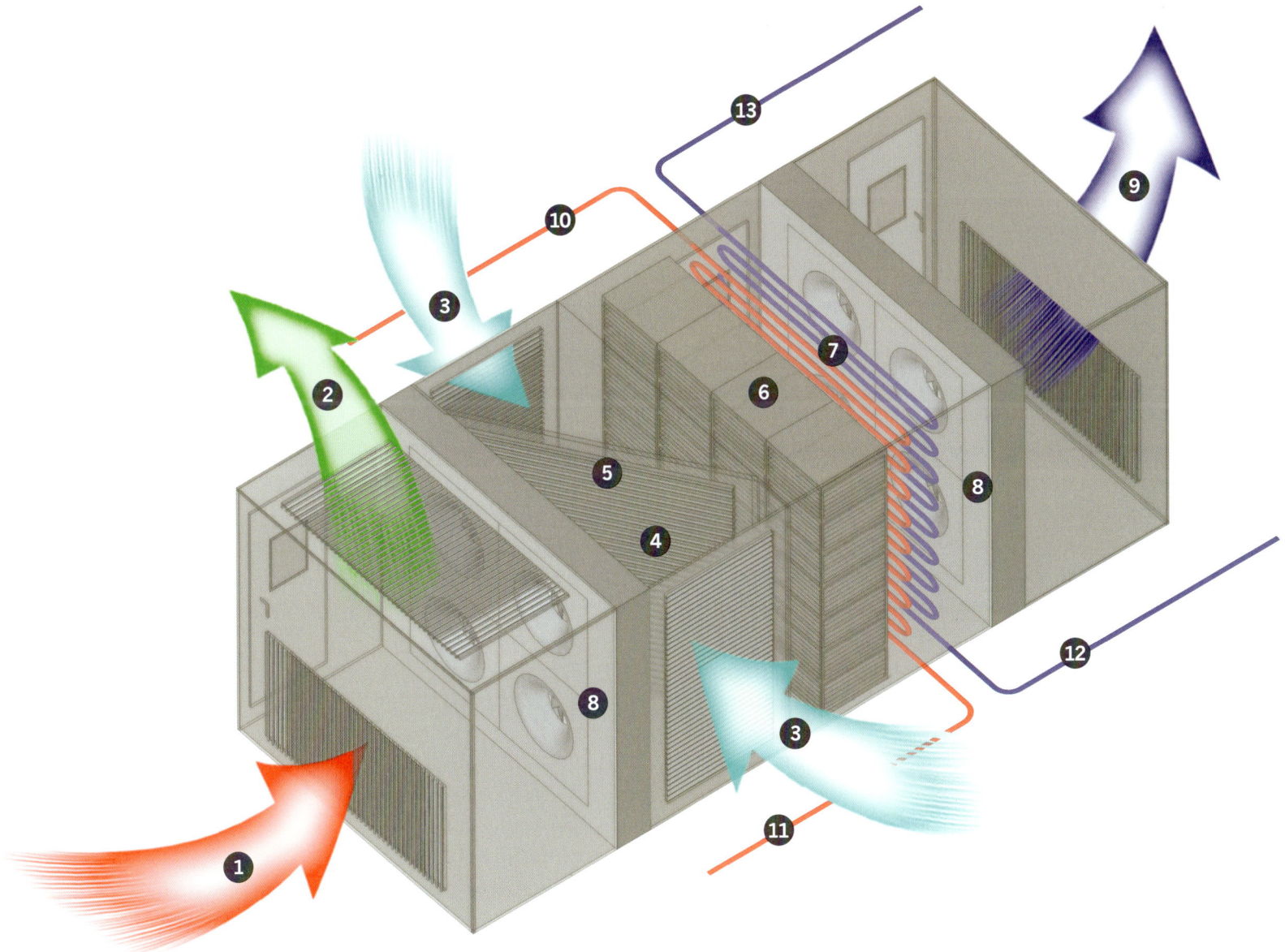

"In 1961–62 the Forest Service made plans to build a road up the beautiful Minam River in Oregon – one of the few roadless valleys in the State. It is choice wilderness – delicate in structure, sparse in timber, and filled with game. We who knew the Minam pleaded against the road. The excuse was cutting timber – a poor excuse because of the thin stand. The real reason was road building on which the lumber company would make a million dollars. The road would be permanent, bringing automobiles in by the thousands and making a shambles of the Minam.

"We spoke to Senator Wayne Morse about the problem...Morse pounded the table and demanded a public hearing. One was reluctantly given. Dozens of people appeared on the designated day in La Grande, Oregon, not a blessed one speaking in favor of the plan. Public opposition was so great that the plan was suffocated."

— *William O. Douglas, Points of Rebellion, New York: Random House, 1969*

INDOOR AIR

A low-velocity underfloor air distribution (UFAD) system supports a majority of spaces in the courthouse, including its six courtrooms. The UFAD provides more efficient air conditioning and better air quality while using less fan power than a conventional overhead ductwork system.

MEP team members also collaborated on new "fan wall" technology within the courthouse. An improved, smaller and acoustically superior type of air handler, the fan wall unit (11 multiple supply and multiple return VAV fan arrays) offers an alternative to large house centrifugal fans to increase fan efficiency and reduce horsepower by decreasing redundancy. Glumac also created a supplemental displacement ventilation scheme to provide air and a portion of the cooling capacity for public spaces. These "air walls" deliver low level, low velocity air down through wall cavities to hidden grilles near floor level. Together with the radiant floor technology, this system reduced ductwork and energy costs while allowing for high-design architectural finishes.

To ensure a healthy indoor air environment, the project team selected low- and no-VOC interior materials, interior paints, shade screening and adhesives. Most of the courthouse flooring is finished concrete to limit off-gassing and the need for waxing and other maintenance. GSA also opted for products that meet South Coast Air Quality Management District rules as well as guidelines created by the Green Seal and Green Label programs. A green cleaning program, specifying use of only nontoxic cleaning products, further reduces indoor air pollutants.

WATER & WASTE

Project design focused on minimizing potable water use within the courthouse through waterless urinals, low-flow toilets, sinks and showerheads, and fixture sensors in the public lavatories. Landscaping around the building is filled with native and drought-tolerant plants, while rain-shutoff sensors and moisture meters further reduce demand for irrigation. The landscape plan also minimized application of insect and disease-controlling chemicals to limit groundwater contamination. These comprehensive low water use strategies combined, reduce the project's total water demand by more than 40 percent versus a comparable, conventional facility.

HEATING & COOLING

Glumac's MEP design features a creative mix of heating and cooling strategies. The UFAD system provides occupants with temperature control in individual work spaces. A central plant utilizes a heat-recovery chiller to reject heat from server room loads into the heating water system. Condensing boilers maximize this system's efficiency while keeping water loop temperatures low to ensure the heat-rejection chiller works efficiently. The building's high albedo roof further reduces the heat island effect.

Radiant technology represents another key design element. A radiant floor system supplies heat to lobbies and public spaces in conjunction with displacement ventilation – particularly important given the building's high ceilings – while offsetting large radiation loads from the extensive glazing. The system consists of PEX tubing encased within a 9-inch (22.86 cm) concrete floor slab, with heated or chilled water from the central boiler and chiller plants circulating through the floor slab. The radiant floor handles 100 percent of the heating load in winter and provides partial cooling in summer.

LIGHTING & DAYLIGHTING

Expansive use of perimeter glazing across the courthouse presented enormous potential for energy savings and an opportunity to reduce dependence on artificial lighting. Although budget constraints initially kept daylighting out of the design, the team continued to search for ways to incorporate it during construction. Incentives from the local electric utility, Eugene Water and Electric Board (EWEB), and tax breaks from the State of Oregon provided the solution. EWEB also contributed funding for the design and integration of controls in the courthouse: automatic-dimming electronic fluorescent lamp ballasts connected to both daylight and occupancy sensors.

Collaborating with GSA and the utility, Glumac's energy modeling determined that daylighting in the building would save an estimated 160,000 kilowatt-hours per year. EWEB provided energy incentives to offset 75 percent of the system's initial cost, with GSA picking up the remainder. GSA also qualified for the State's Business Energy Tax Credit (BETC) program. This collaboration resulted, after the rebate, in a minimal net cost for the daylighting system.

SITE

The site's landscaping plan connects the courthouse with open space to the east and the Willamette River and a park to the north. Native plants continue to thrive in Eugene's cool, wet climate, requiring minimal irrigation and no pesticides, while incense cedar, quaking aspen and European beech populate the area between the courthouse and a bordering street. These plantings also provide natural windbreaks against prevailing winds from the west, sound dampening and exhaust mitigation from adjacent traffic, and sun shading.

Additional low-impact development strategies on site include a limited parking area (plus below-grade parking with space for eighty cars) and the avoidance of contiguous, impermeable surfaces to reduce storm water runoff. Since its installation in 2006, the landscaping has restored approximately 37 percent of the site's soils to natural, pre-development percolation rates.

INTEGRATING HVAC AND DAYLIGHT

A FINE BALANCE: Cross-section shows the courthouse lobby and circulation spaces – featuring daylight and radiant heating and cooling – a design which highlights the building's raw interior materials and keeps all public areas open.

1　Entry lobby, atrium and circulation space with radiant slab heating and cooling
2　Office space with underfloor air distribution system and daylighting controls
3　Atrium skylights
4　Underground parking uses the HVAC system's exhaust to temper the space
5　Underground parking garage door doubles as HVAC relief grille

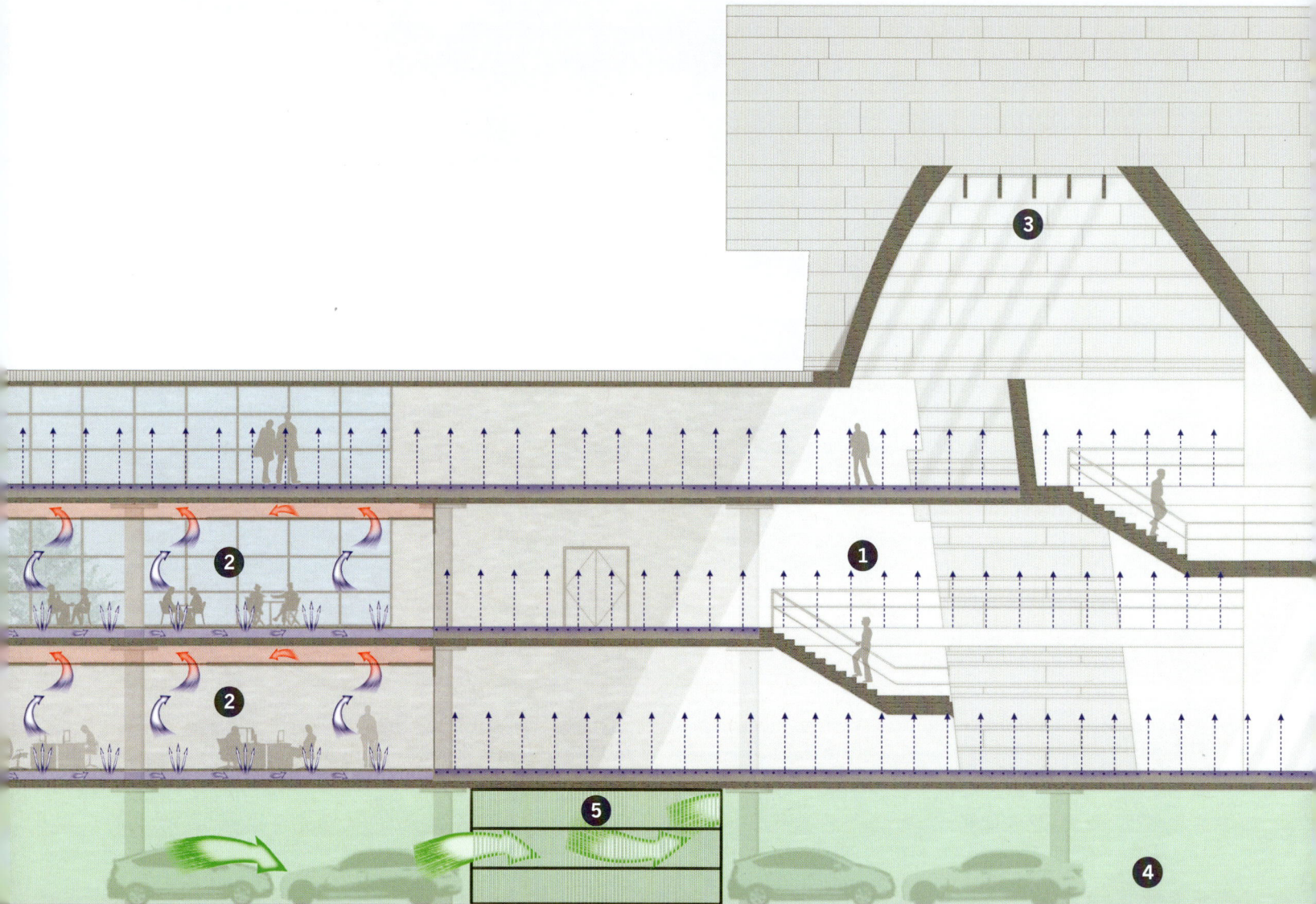

AIR CIRCULATION

1 Fresh air intake shaft
2 Fresh air plenum/mechanical room
3 Fan wall air handlers
4 Exhaust discharge into underground parking
5 Underground parking uses the HVAC systems exhaust to temper the space
6 Underground parking garage door doubles as HVAC exhaust grille
7 Supply air
8 Underfloor air distribution system
9 Independent mechanical system for courtroom pods and judges' chambers

10 Independent mechanical system for perimeter circulation space
11 Return air
12 Office space with underfloor air distribution system
13 Return air plenum in suspended ceiling

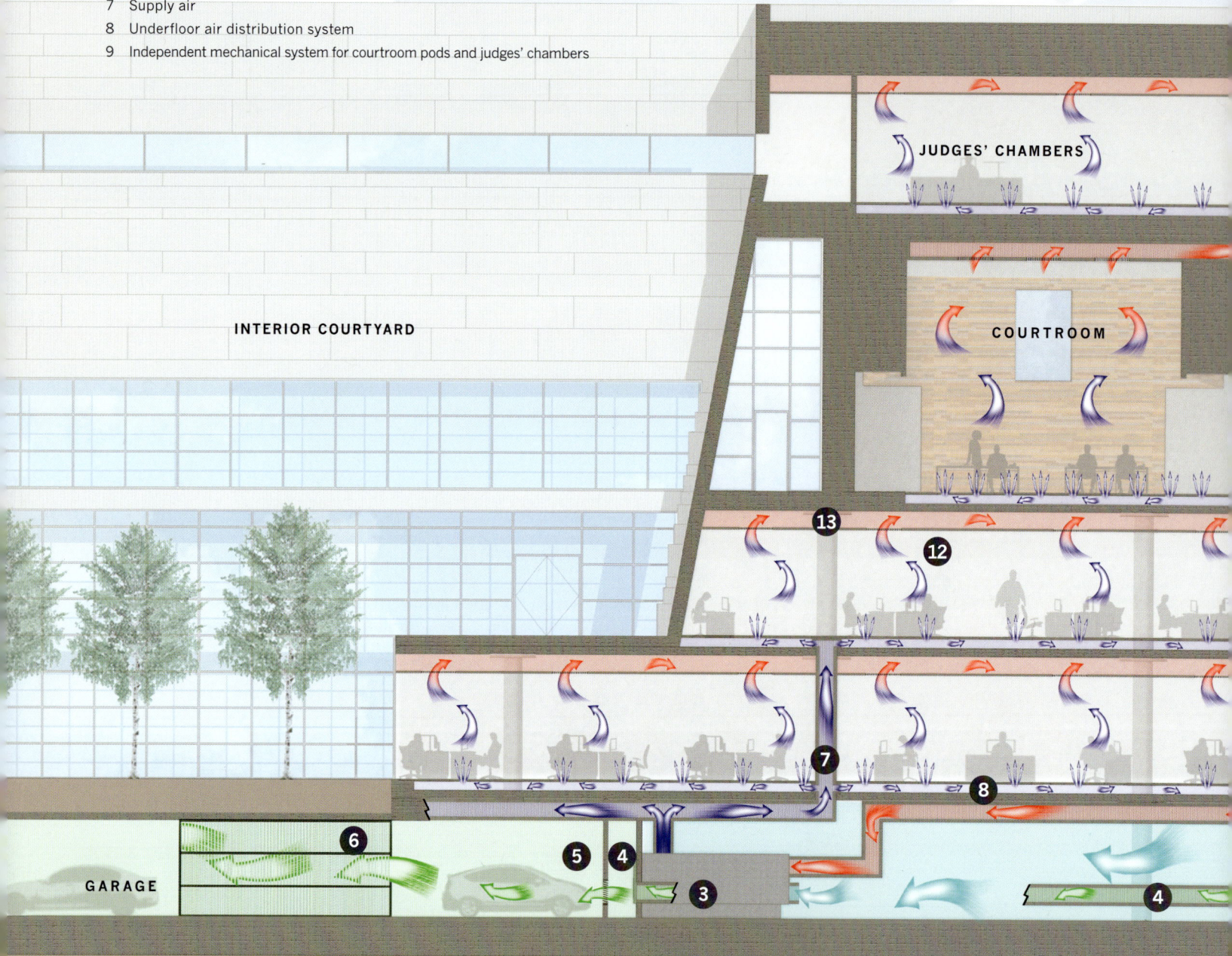

JUDGES' CHAMBERS

COURTROOM

INTERIOR COURTYARD

GARAGE

AND FRESH AIR FOR ALL: An integrated design process and extensive use of Building Information Modeling (BIM) tools, allowed members of the design team to arrive at collaborative mechanical solutions and rethink the use of judicial spaces throughout the courthouse.

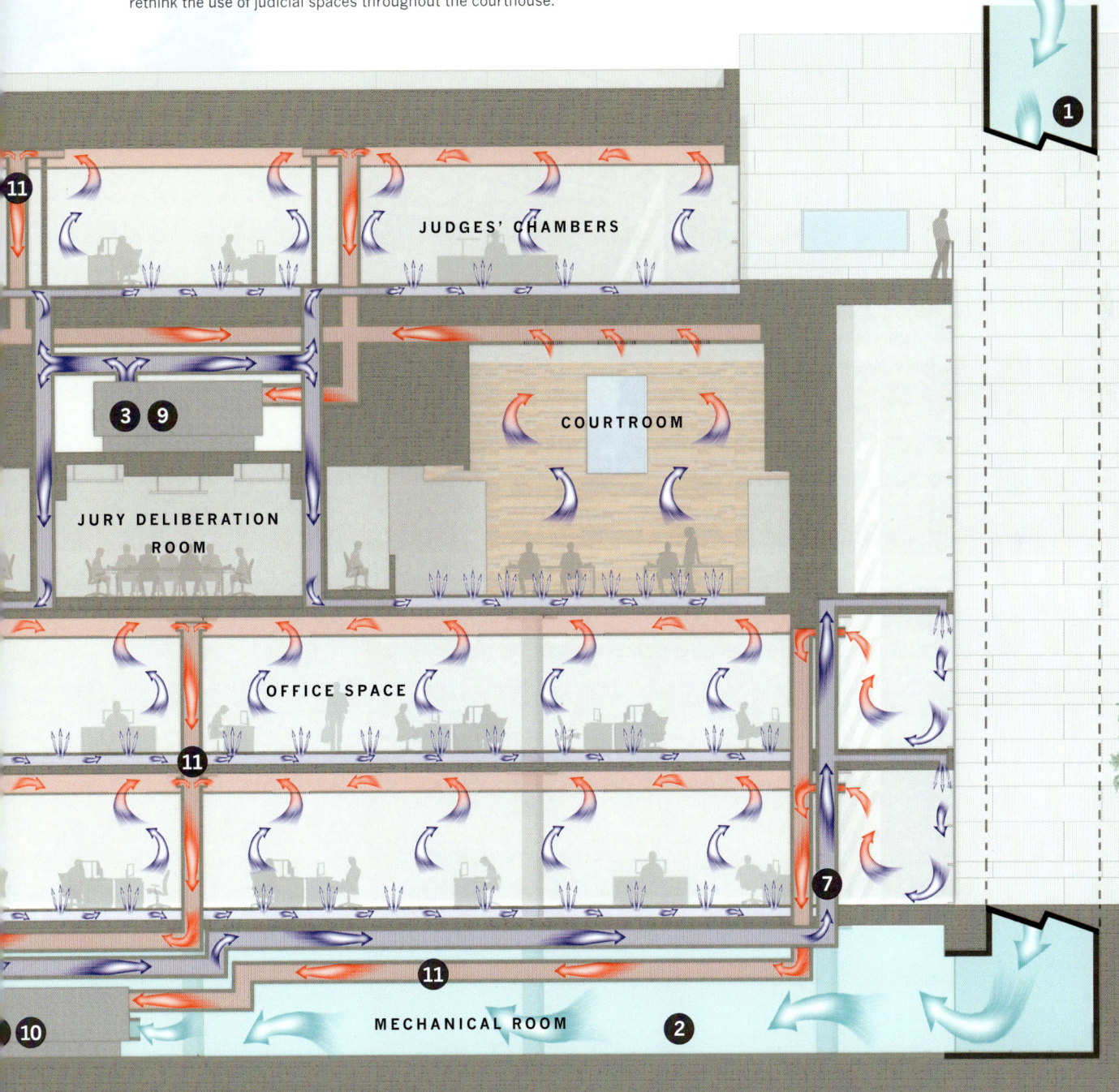

1

11

JUDGES' CHAMBERS

3 **9**

COURTROOM

JURY DELIBERATION ROOM

OFFICE SPACE

11

7

10

11

MECHANICAL ROOM

2

THE RULING: Unprecedented for courthouse design, courtroom daylighting introduces natural light from adjoining circulation areas while addressing security requirements.

ENERGY

Through use of energy modeling software (eQuest version 3.54), Glumac's designers generated vital information on first costs, lifecycle costs, energy and equipment upgrade costs, as well as various design and equipment options. As built, the courthouse has yielded energy use reductions of approximately 39 percent over a comparable baseline model, ASHRAE Standard 90.1. To achieve this efficiency, the building incorporates an impressive array of green strategies. Large exterior windows and high ceilings increase daylighting for energy efficiency; the high-performance glazing also prevents heat loss and unwanted solar heat gain. Exterior sunscreens on the lower two floors manage solar cooling loads, while shading structures shelter southern windows from unwanted solar heat gain.

Also important to the design scheme, all indoor air is purged from the building at night and replaced with ambient air, thereby reducing cooling loads. A displacement air delivery system enabled a smaller mechanical system overall. To further improve energy savings, designers specified high-efficiency, variable-speed motors – sized for aggregate load averages rather than peak loads.

Finally, the project team recommended spending an additional $470,000 to upgrade various mechanical system features – a decision that would result in $42,000 in energy savings yearly and an 11-year simple payback. By applying for more than $300,000 in utility incentives, based on exceeding stringent state codes, the client reduced this payback figure to four years.

COMMISSIONING

Glumac led commissioning efforts throughout the design and construction phases. Tasks included oversight and verification of the commissioning plan, review of the contractor's submissions, production of a commissioning manual for ongoing operations, and submittals of contracts for ongoing reviews during the first year of occupation.

Again, Glumac and GSA worked together in developing a procedure for testing air tightness to ensure the integrity of the raised-access floor plenums. The initial test on a 4,000-square foot (370-square meter) mock-up area indicated an excess of 70 percent leakage. Utilizing smoke machines, the leaks were located and repaired. Project goals overall established a target of less than 10 percent leakage for all floor plenums.

"Results like Toyota's are helping to spark a budding "green revolution" in American workplaces. The movement is starting to change how office buildings are designed and could render thousands of existing offices obsolete."

— *Roger Vincent, "The Greening of Work," Los Angeles Times, August 27, 2006*

TOYOTA MOTOR CORPORATION

CUSTOMER SERVICE / FINANCIAL SERVICES CENTER HEADQUARTERS
TORRANCE, CALIFORNIA

Proof of the economics of green design is on display at Toyota's Southern California headquarters. Together with LPA Inc., Glumac added a state-of-the-art work environment to the company's existing campus, while integrating a number of sustainable design elements at a cost highly competitive with spec office development. In addition to its high-performance thermal envelope, the building features a 2,700-ton (9,500kW) direct-fired natural gas absorption chiller plant, one of the first applications in Southern California using reclaimed water for its cooling tower makeup. Excellent indoor air quality, a priority for Toyota, relies on 95 percent efficient air filtration systems and CO_2 sensors. Energy efficiency exceeds state code requirements by more than 20 percent, and together with a photovoltaic roof array, the project uses 50 percent less energy from the state's electric grid than a standard building.

"This proved to the world that you could build a very efficient, LEED certified office for similar dollars per square foot to any developer-led suburban office building. It was really the cornerstone of a cost-efficient, sustainable design – and at the time, became the largest LEED Gold project in the country."

— *Richard Holzer, Principal, Glumac*

PROJECT DETAILS

Three-story, 625,000-square foot (58,000-square meter) office building

Completed 2003

Architect: LPA

Contractor: Turner Construction

LEED Gold certified

"Providence Health System demonstrated a stronger push toward a sustainable healthcare facility than we'd ever seen before. And their definition of sustainability was not limited to just energy efficiency. It included daylighting, water efficiency, lighting efficiency, heat recovery, and a unique natural ventilation system to improve indoor air quality – all designed to promote a greater healing environment for the main occupants of the building — the patients."

— *Leonard Klein, Principal, Glumac*

PROVIDENCE NEWBERG MEDICAL CENTER

NEWBERG, OREGON

With care and foresight in creating a healthy environment for patients and staff, the Providence Newberg Medical Center became the first LEED Gold hospital in the United States. Fresh air, green power and high-efficiency mechanical systems all combine within the Center to reduce energy costs and ensure optimum indoor air quality. This new outpatient facility also includes forty inpatient beds, fifteen short-stay beds and four operating rooms, and room for future growth. Additional high-performance building features include: HVAC systems that employ 100 percent outside air/100 percent exhaust to enhance infection control, condensing boilers operating at 95 percent efficiency, premium efficiency chillers which, in combination with low-flow fixtures, reduce water usage by more than 20 percent dispatchable standby generation and centralized lighting control systems with occupancy sensors and daylight controls throughout.

"Green building fits well with Providence's core values of respect, compassion, justice, excellence and stewardship..."

— *Amy Eagle, "Gold star,"* Health Facilities Management, *July 2007*

PROJECT DETAILS

Three-story, 143,000-square foot (13,300-square meter) hospital, 43,000-square foot (13,300-square meter) medical office building

Completed June 2006

Architect: Mahlum Architects

Contractor: Skanska

LEED Gold certified

TWELVE|WEST, 2009: Rising
22 stories over Portland's
West End, the 85,000-square
foot (7,897-square meter)
mixed-used building sets a new
tone for revitalizing its historic
neighborhood near downtown.

CHAPTER EIGHT:
TWELVE|WEST

GREEN TOWER RISING

"There should be different places in the sky for socializing."

— *Kenneth Yeang*
Big & Green: Toward Sustainable Architecture in the 21st Century
p.176, Gissen, David, Editor, New York: Princeton Architectural Press, 2002

"The building itself has beautiful color even when the skies are gray – and remains bright and lively across the seasons. Even in Portland's often overcast environment, Twelve|West's design and use of glass makes it possible for a lot of sun to enter, even as we're seeing the changes of the day and the movement of light outside like never before."

— Gene Sandoval
Lead Designer, Zimmer Gunsul Frasca

PORTLAND, OREGON

In a city filled with sustainably-designed buildings, topped by green roofs, harvesting the rain and sunlight, it stands out – most immediately as one of the first urban buildings in the nation to integrate small-scale wind energy within its 22-story design. Beyond its impressive 40-foot-tall (12.2-meter) wind turbines overhead, this new mixed-use high-rise also makes prominent use of storm water management, high-efficiency radiant heating and cooling, natural ventilation, and a rich variety of recycled and reclaimed materials. Yet Twelve|West is still more: a symbol, a statement even, of urban livability, of elegantly connecting past and present, the promise of things to come.

The building's origins date back to 2005. Zimmer Gunsul Frasca Architects LLP (ZGF) needed more space to accommodate its growing workforce, which specializes in architecture, planning and interiors. Recognized with more than four hundred national, regional and local awards, the firm has been an early leader in sustainable design – completing more than twenty-five LEED buildings across the U.S. At the time housed within a renovated, five-story warehouse in the heart of Portland's Old Town (since the late 1970s), ZGF's leadership considered relocating to an emerging neighborhood where its new headquarters office might serve as a catalyst for growth and development.

The West End section of downtown presented one intriguing possibility. Wedged between the acclaimed Pearl District and Portland's Central Business District, this mature neighborhood is home to a mix of churches, grocery stores, schools, high-density residential buildings and restaurants as well as 24/7 nightclubs and shelters. The West End also features a large stock of historic properties prime for renovation. Many of these buildings date back to the early 1900s and include single-room occupancy hotels constructed for the Lewis & Clark Pavilion to house men working on the railroads and lumber mills then lining Portland's waterfront.

OPEN DOOR: Occupying four floors of the new tower, Zimmer Gunsul Frasca Architects LLP (ZGF) wanted a place to showcase sustainable design – while fostering innovation and interaction among associates as part of its open office plan.

Within this urban landscape of seven blocks, comprising a large number of surface parking lots, ZGF saw great potential to create a meaningful connection from the pedestrian-friendly Pearl to the traditional downtown core and mixed-use neighborhoods further to the north and west. So did Gerding Edlen, developer of several LEED projects in the adjacent Pearl District, including the 1.7-million-square foot (158,000-square meter) Brewery Blocks. "We envisioned moving sustainable development across Burnside," said Mark Edlen, the firm's managing principal, "to extend the energy and street life of the Brewery Blocks to the West End as well." Enter the Goodman family, the third element of this successful alliance. The Goodmans own a large number of parking lots in downtown Portland, with a desire to see many of those properties converted to highly sustainable, mixed-used buildings over time.

At 85,000-square feet (37,200-square meters) and occupying roughly a half block between SW 12th and 13th Avenues and Washington Street, ZGF's new office would be a beginning, one of the first significant new buildings to take up residence in Portland's West End. Twelve| West achieved two LEED Platinum certifications: one for New Construction, another for Commercial Interiors. On July 14, 2009, ZGF's 250 Portland employees took occupancy of their stunning new space over four floors of the building. Combined with market-rate apartments (available for leasing that October) and a vibrant new work environment,

Twelve|West promises to bring more people and economic and cultural vitality into the city's core.

ENVIRONMENTAL OBJECTIVES

In addition to defining a fundamental role in urban renewal for the area, ZGF further framed the project's chief goals with the 2030 Challenge: a worldwide, voluntary initiative to reduce fossil fuel emissions and energy consumption through sustainable design strategies for new buildings, developments and major renovations. Gerding Edlen, Glumac and the architecture firm had all joined the 2030 Challenge and agreed to adopt its performance targets for Twelve|West.

Glumac's designers posed seventy different energy and carbon reduction strategies for the building, which were later narrowed to twenty-five possible solutions. Modeling predicted that select energy efficiency strategies would reduce consumption of energy by more than 44 percent beyond ASHRAE 90.1-2004 and exceed the 2030 Challenge benchmarks for this project type.

DESIGN OVERVIEW

Throughout the building's concept and design phases, ZGF was clear about the purpose of its move to a new location. "We wanted to be on a site that would help revitalize a re-emerging neighborhood," observed Gene Sandoval, a ZGF partner and lead designer for the project. "That brings with it the responsibility

"Its architecture is contemporary – and stunning – and yet the design of the skin and how they put the different components of the building together is equally exciting to me, both for the exterior and interior. Twelve|West is a pioneering project within an emerging neighborhood."

— *Mark Edlen, Gerding Edlen*

to achieve a balance between live and work, the whole idea of mixed use. We also wanted to combine rental housing with an architectural office and retail – and do it well as a 'hybrid' building." As the building owner and anchor tenant, the firm occupies the second through fifth floors of Twelve|West, with street-level retail spaces and five levels of below-grade parking plus studio, one-, two-, and three-bedroom apartments spread over the upper seventeen floors.

With five offices coast to coast and a growing portfolio of LEED projects, ZGF intended the new Portland headquarters office to serve as a living model to inform future sustainable design. In addition to the four urban wind turbines overhead, the building's eco-roof (rooftop garden and terrace space) diverts and cleans rainwater while significantly reducing roof temperatures in warmer months. Designers also ensured the tower maximizes natural light by choosing a south-facing site, its transparent façade both embracing and reflecting the sun; in winter, the dull rainy sky casts the curtain wall glass a steely shade of blue.

The design process afforded ZGF a unique opportunity to create its new space from the ground up, beginning with an office-wide "culture charrette" that asked participants: "How do we increase interaction and make the most of public space? How can the office foster further innovation?" Feedback on workstations and workflow then led to employees brainstorming as teams to plan and design

an open interior space for the firm, with room to collaborate on projects and improve overall productivity and quality.

After selecting its project team, including Hoffman Construction as construction manager and Glumac for the MEP systems, ZGF conducted a series of design charrettes to further define several key objectives that would inform the sustainable design of Twelve|West. These objectives included:

» Create a structure that carefully unites the live/work/learn/ play components of the building
» Construct a transparent building – inviting and active – that connects the building's inhabitants to the urban landscape, while taking advantage of natural light
» Ground the building in a manner that promotes active street life
» Accommodate exterior gardens and terraces for areas of interaction as well as respite
» Integrate advanced and symbiotic sustainable building systems to promote natural resource conservation
» Design homes for lease that maximize space, light, views and include luxurious amenities and energy savings

Designers also gave careful consideration to the large number of historic properties nearby in the West End – and the impact of future new construction throughout the neighborhood. Despite

CREATING A LEED CULTURE: ZGF's Portland office integrates thematic public art with active technologies (high-efficiency lighting controls, underfloor air) and passive design elements (daylight harvesting, chilled beams) – another opportunity to highlight its growing portfolio of LEED projects.

BLUE SKY THINKING:
Reaching into the Portland
sky, the highly transparent
façade admits controlled
daylight to more than 90
percent of regularly occupied
spaces within the office floors
and residential units.

"One percent and no payback is a fine start, because Twelve|West is more R&D, more artist's concept sketch, than replicable business model. It is a bold leap into the gap. Twelve|West wind exists to demonstrate technical feasibility, to stoke thought and provide lessons for those who follow, more so than to power the building."

— *Chris Davis*
"Launching Into the Urban Wind Gap," PowrTalk, Discovery Channel, November 27, 2008

the challenges of a constrained building site, all three adjacent historic structures were preserved during construction.

The primary program elements within Twelve|West – ZGF's Portland office and the 274 rental apartments – further reflect a wide palette of sustainable materials, interior finishes and design choices. Throughout the office, salvaged wood as well as high-recycled content and urea-formaldehyde-free MDF cores were selected for doors, casework and painted trim, in addition to recycled blue jeans for insulation and 96 percent recycled, locally-manufactured gypsum wallboard. To enhance the open floor concept, several interior offices feature transparent glass walls so natural light penetrates deep into the building. Collaboration throughout ZGF's design practice is achieved by: custom-designed workstations, thirteen conference rooms, a daylit resource library, and "social atriums" (wide stairways) to connect various floors.

The "Indigo @ Twelve|West" apartments begin on the building's 6th floor, featuring a range of units from 541- to 2,125-square feet (50- to 197-square meters) and three floors of penthouse homes. ZGF specifically chose interior materials and colors to optimize the daylight entering each apartment. Concrete was left exposed at ceilings and walls where possible. To maximize natural light, designers oriented living and bedroom spaces along the floor-to-ceiling glass curtain wall of the building;

using operable windows and balconies allow natural ventilation. Sustainable finish materials include bamboo veneer casework in kitchens and bathrooms, carpet in bedrooms with 25 percent recycled content, granite counter tops in kitchen and baths, PVC-free window coverings, and zero-VOC paints.

MEP DESIGN PROCESS

The 2030 Challenge and a number of LEED goals all set high expectations for the project to pursue innovative, often breakthrough, designs to maximize daylight and views, minimize solar gain, and achieve healthy living and work environments throughout the tower. "Especially because it's our own building," noted Sandoval, "we were willing to take some calculated risks to achieve something special and yet knew we needed to partner with designers also willing to take these systems to the next level."

In particular, ZGF pressed designers to make mechanical and plumbing systems as green as possible, while ensuring comfort and flexibility since individuals and studios move around the office frequently. The team examined multiple schemes for night flushing, natural ventilation, and radiant cooling and heating. Early in schematic design, ZGF also challenged Glumac to push overall lighting levels down for occupied spaces and emphasize natural light. This approach resulted in the use of LED path lighting at desktop surfaces, unconventional lighting controls and other strategies to dramatically reduce electric lighting loads.

Glumac's engineers devised an energy matrix specifically for this project. The new tool – comparing various options to meet aggressive 2030 Challenge and LEED targets – became instrumental in convincing ZGF to consider underfloor air and chilled beams within the building. "Historically, a lot of energy analysis happens separately from design development," explained Leonard Klein, Principal with Glumac. "For this project, we deviated from the tried and true stacks of detailed energy analyses by simplifying everything into an 11 by 17-inch (27.9 by 43.2 cm) spreadsheet that summarized all the pros and cons, energy requirements, CO_2 impact, and overall construction cost."

Using multiple CFD simulations to model the dynamics of underfloor air distribution and overhead chilled beams at the office perimeter, Glumac designers could project building performance under different time-of-day conditions: from early morning to peak load. More specifically, the CFD analyses compared how systems would respond on south-facing versus east-facing and west-facing surfaces and then for the large open layout and within enclosed offices.

As with initial concepting and the charrettes, collaboration was key – and unique – across the entire team, particularly for what was largely a design-assist project after the DD phase. For Glumac, working alongside subcontractors early in MEP design led to fewer conflicts and a streamlined design process overall.

INDOOR AIR

Natural ventilation design for Twelve|West focused on the use of operable windows and an underfloor air distribution (UFAD) system. Glumac's analysis of operable window areas for the office floors and apartments also assisted ZGF designers with optimal placement of windows for proper airflow.

Nightflushing is also essential to building performance, particularly on the office floors. To accomplish this effect, Glumac designers specified fanwall technology. This array of plenum fans produces lower noise levels, minimizes static pressure and improves maintenance through built-in redundancies, therefore leading to bigger energy savings. Air handling units also feature an economizer mode, adjustable from minimal to 100 percent outside air according to the temperature of supply air and time of day. In addition, variable speed drives reduce airflow, allowing air to slowly meander through the building.

WATER & WASTE

The building's comprehensive approach to water efficiency includes conservation measures, storm water reuse and solar water heating. All office floors, apartments and the penthouse homes feature Energy Star appliances and low-flow plumbing fixtures which, together, reduce water use by more than 44 percent.

SOCIAL SPACES: Available to residents and ZGF staff, the rooftop garden and terrace spaces next to the building's green roof offer unparalleled views of Portland's Pearl District and the downtown neighborhoods.

"All along, it was important to discuss how we would accomplish the different LEED targets with mechanical, electrical and plumbing systems. And each deliverable we hit would be re-evaluated: Are we still in line to meet these goals? What do we need to change to keep us on target? Without question, it remained a collaborative effort throughout for the architects and engineers."

— *Jennifer Streb, Glumac*

Central to the design is its rainwater harvesting system, which gathers and diverts rooftop water to a 50,000-gallon (189,300-liter) storage tank in the building's underground garage for filtration and treatment (UV) prior to reuse. This system collects approximately 273,000 gallons (1,033,300 liters) of rainwater annually, in addition to 13,000 gallons (49,200 liters) of condensation water from air handlers during the summer months. Consequently, the entire water/wastewater scheme meets 100 percent of the green roof's irrigation needs and 90 percent of toilet flushing demands for the office.

This projected reduction of the building's combined sewer totals led directly to a 30 percent cut in the Systems Development Charge (SDC) from the City of Portland Bureau of Environmental Services. A savings of $204,840 covered 91 percent of the first cost of the system, vastly reducing the simple payback period for this investment.

HEATING & COOLING

Left exposed on the interior where possible, the concrete structure of the building provides thermal mass to moderate indoor air temperatures. This mass cools with night air in the summer months and absorbs excess heat throughout the day.

The biggest source of office cooling for Twelve|West, though, lies in the combination of underfloor air and natural ventilation through operable windows. The UFAD system delivers air directly to occupied zones at floor level for more moderate temperatures and velocities than a conventional, mechanical ventilation system, using less energy and stressing individual comfort. Employees also gain personal control of air flow via adjustable diffusers at each workstation. Overhead, apartments feature traditional fan coil units with chilled water cooling, each incorporating very high-efficiency motors with pre-programmed operating speeds. For perimeter heating, apartments also include underfloor electric baseboard units, whereas the retail and office floors rely on a heating/hot water loop with condensing boilers (low temperature, high efficiency).

To further enhance perimeter cooling, designers placed passive chilled beams throughout the four office floors. In exploring the chilled beam concept with the architect, Glumac considered active and passive schemes – both designed to add energy-efficient cooling capacity on particularly hot days. Ultimately, the team settled on passive technology, which saves energy by moving heat with water instead of air and moving air without the use of fan energy. ZGF selected an equipment supplier not yet represented in the United States; as a result, a series of chilled beams were sized and custom fabricated to fit specific areas. These chilled beams, each 3-feet by 10- or 12-feet (0.9 meters by 3.0 meters or 3.6 meters) in dimension, produce a 65 percent radiant/35

URBAN WIND: Four 12-foot (3.66 meter)
diameter, building-mounted turbines generate
10,000 to 12,000 kWh annually – enough to
power the building's core while serving as a
visible demonstration of urban energy potential.

SUSTAINABLE VISION

Designed to meet two LEED Platinum Certifications and serve as a laboratory for cutting-edge, sustainable design strategies, Twelve|West incorporates a broad range of sustainable features – from rainwater reuse and daylighting to state-of-art energy controls – as one of the world's first building integrated wind turbine projects.

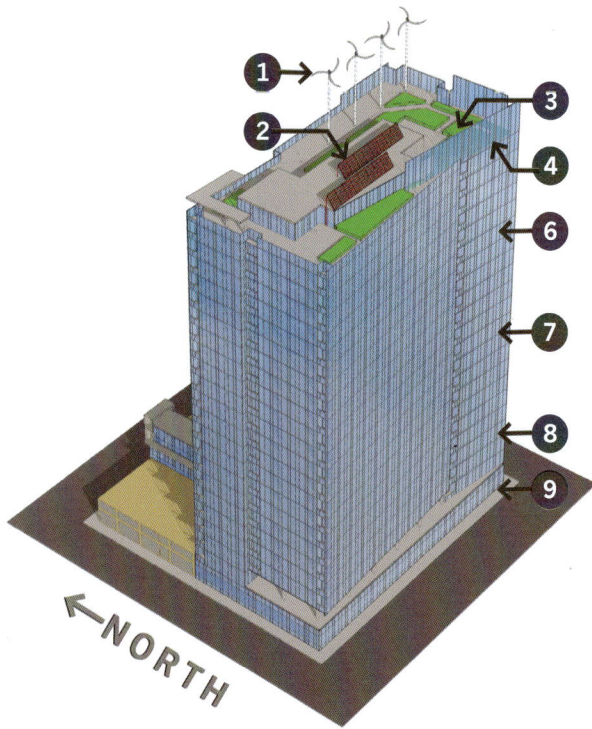

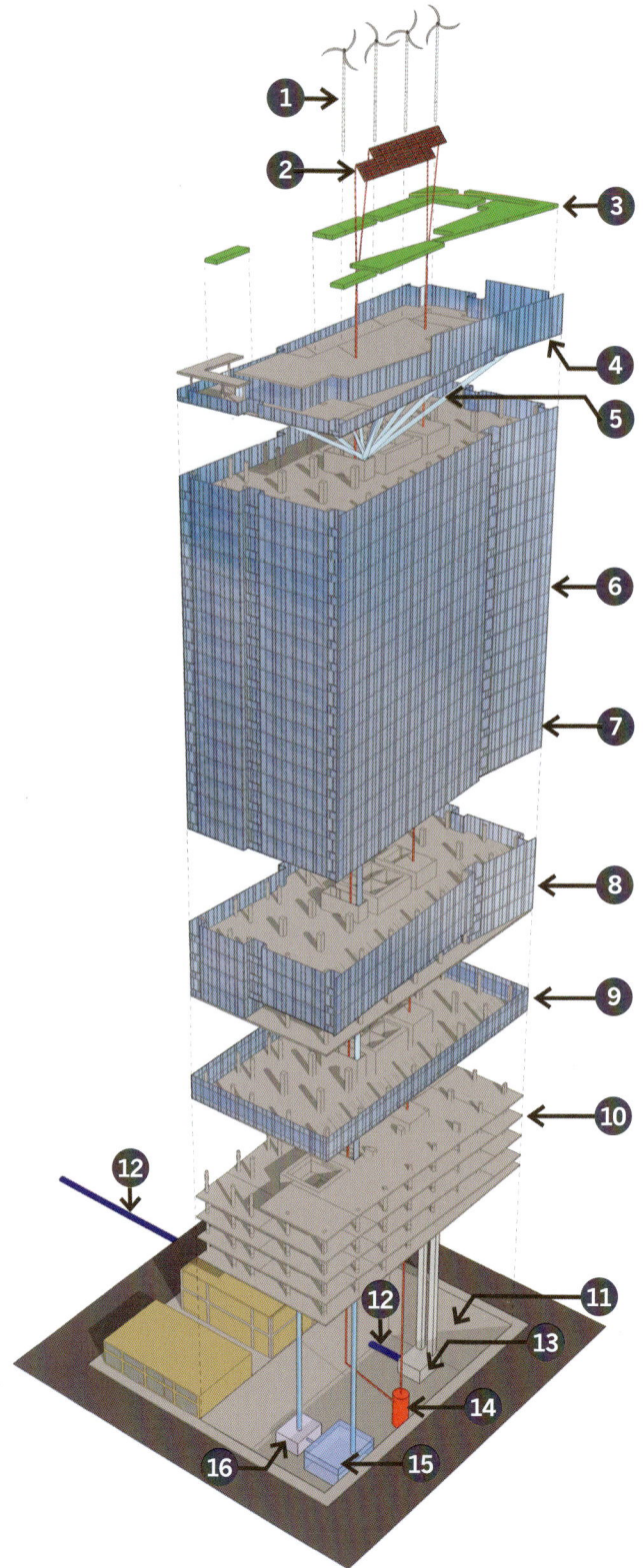

← NORTH

1 Four rooftop wind turbines produce between 10,000 and 12,000 kWh of electricity per year.

2 Solar thermal panels provide 24 percent of the building's hot water. The system lowers the building's natural gas consumption, which lowers both the economic cost to the owner and the ecological impact of gas emissions.

3 Roof gardens collect and filter some of the rainwater that falls on the building and reduce the roof's temperature during summer months.

4 The entire roof is used as a rainwater collector. Roof drains with natural filters (using compressed leaf material) clean the water as it enters the reclamation system.

5 Below the roof, water is consolidated and piped to the basement for further filtration and storage.

6 Seventeen floors of studio, one, two, and three bedroom apartments

7 Glazing: Low-e glass provides an optimal ratio of visual light transmittance and minimal solar heat gain, allowing the building to have floor to ceiling glass throughout. The glass admits 35 percent of the visual light while reflecting 74 percent of the associated heat, lowering energy use for lighting and cooling.

8 ZGF Architects' office occupies the second through fifth floors.

9 Retail stores and lobby space for the INDIGO @ Twelve|West residences and ZGF Architects' office above, occupy the ground floor.

10 Five floors of below grade parking

11 Mechanical space for rainwater treatment and storage is located on the lowest level of the basement.

12 Chilled water is supplied by a central cooling plant located in the nearby Brewery Blocks district.

13 Air handlers

14 Water heater works in unison with the rooftop solar thermal collectors to maximize efficiency.

15 Rainwater storage tank can hold up to 23,000 gallons (87,100 liters).

16 Ultraviolet filters clean rainwater before it is pumped to the office levels of the building for use in flushing urinals and toilets.

> "Creativity begins with an idea – seeing things differently."
>
> — *Jun Kaneko*

percent convective effect, offering greater surface area exposed to occupants and more comfort than traditional models.

Finally, Glumac needed to locate a source of chilled water for space cooling. Chillers on the roof of a nearby building in the Brewery Blocks development in the Pearl District offered one possibility. Eventually, ZGF and Gerding Edlen were able to negotiate a contract to run utilities across Burnside Street to Twelve|West. The result: essentially, a district plant to share energy-efficient cooling loads between the two buildings.

LIGHTING & DAYLIGHTING

Twelve|West's interiors celebrate light. Glumac focused on multiple energy-saving strategies to emphasize daylight harvesting and demand response, occupancy controls to avoid lighting unused spaces and automated shades to maximize natural light usage. As built, more than 90 percent of all regularly occupied spaces within the office floors and residential units are now daylit. In this highly transparent tower, however, designers also had to take great care in addressing solar gain along the building's south façade without losing access to the views outside. Here, they selected low-e vision glass, a solution which admits 35 percent of visible sunlight but reflects 74 percent of the associated heat, reducing energy use for lighting and space cooling.

With input from Gene Sandoval and his team, Glumac's designers ran morning-to-evening simulations to calculate daylight projections for each space and establish the building-wide control scheme. Primarily, ZGF wanted the ability to fine-tune light levels in areas with ample daylight and establish thematic pre-sets for special functions. After reviewing a number of options, they selected Lutron Quantum – a DALI (Digitally Addressable Lighting Interface) protocol adopted in many European designs but still relatively new in the U.S. – as the centralized control system for occupancy, daylighting, shades and demand response. In addition to integrating architectural and energy management controls, the system offers precise control of every lighting fixture, providing the firm with greater flexibility to adjust lighting levels as they modify how spaces are used within the office.

SITE

Nestled within Portland's urban core, construction of Twelve|West transformed an existing surface parking lot and low-rise commercial building into a site with storm water attenuation and habitat potential.

More than 4,000-square feet (370-square meters) of the building roof features a variety of native and adaptive plants, effectively reducing both storm water runoff and the building's contribution to the urban heat island effect while providing a green amenity for ZGF employees and residents. This mix of evergreen, deciduous plants and seasonal flowering bulbs and grasses on the roof uses 85 percent less water than comparable landscapes; where irrigation is needed, rainwater collection plus condensation from the mechanical units feeds a drip-line emitter system.

ENERGY

Wind – utilizing four building-mounted turbines – is expected to generate 10,000 to 12,000 kWh per year, approximately 1 percent of energy requirements, for Twelve|West and enough to power its elevators. Additionally, solar thermal panels heat 24 percent of hot water used in the building, offsetting natural gas use.

Although the wind turbines contribute only incrementally to building performance, ZGF, Gerding Edlen and the rest of the team believe their inclusion meets the progressive goals of the project while advancing research on future urban applications for wind. An extensive study of local wind patterns concluded the idea would work if turbines were placed high enough above the roof's 279-foot (85-meter) elevation to catch prevailing winds, which flow across the city from the northwest in summer and the opposite direction in winter. Just as important, instrumentation to monitor actual turbine performance and wind flow patterns will be used to validate predictions. Recognizing the ground-breaking nature of this demonstration project, the Energy Trust of Oregon and the Oregon Department of Energy underwrote the entire system cost through energy efficiency grants and tax credits.

COMMISSIONING

In addition to MEP design, Glumac performed all commissioning on the building – a process integrated with both the design/construction and review of sustainable energy features.

1. Direct sunlight
2. Indirect (sky dome light)
3. Reflected light
4. Photo cells (light sensors)
5. Lights (controlled by building automation system)
6. Sun shade (controlled manually)

SUMMER/WINTER SUN: Glumac designers calculated multiple daylighting scenarios within the ZGF office spaces. As a result, the building takes advantage of daylighting during the majority of working hours, allowing electric lights to be turned off during the day. When the sun is low in the sky (early morning, late afternoon and winter months), causing glare, employees can manually lower shades. In addition, photocells sense light levels have dropped and will automatically activate the lights.

OVERHEAD RADIANT EFFECT: Passive chilled beams, each measuring 3-feet wide by 10- or 12-feet (0.9 meters by 3.0 meters or 3.6 meters), produce a 65 percent radiant/35 percent convective effect within ZGF office areas, resulting in more exposed surface area and greater occupant comfort.

DAYLIGHTING

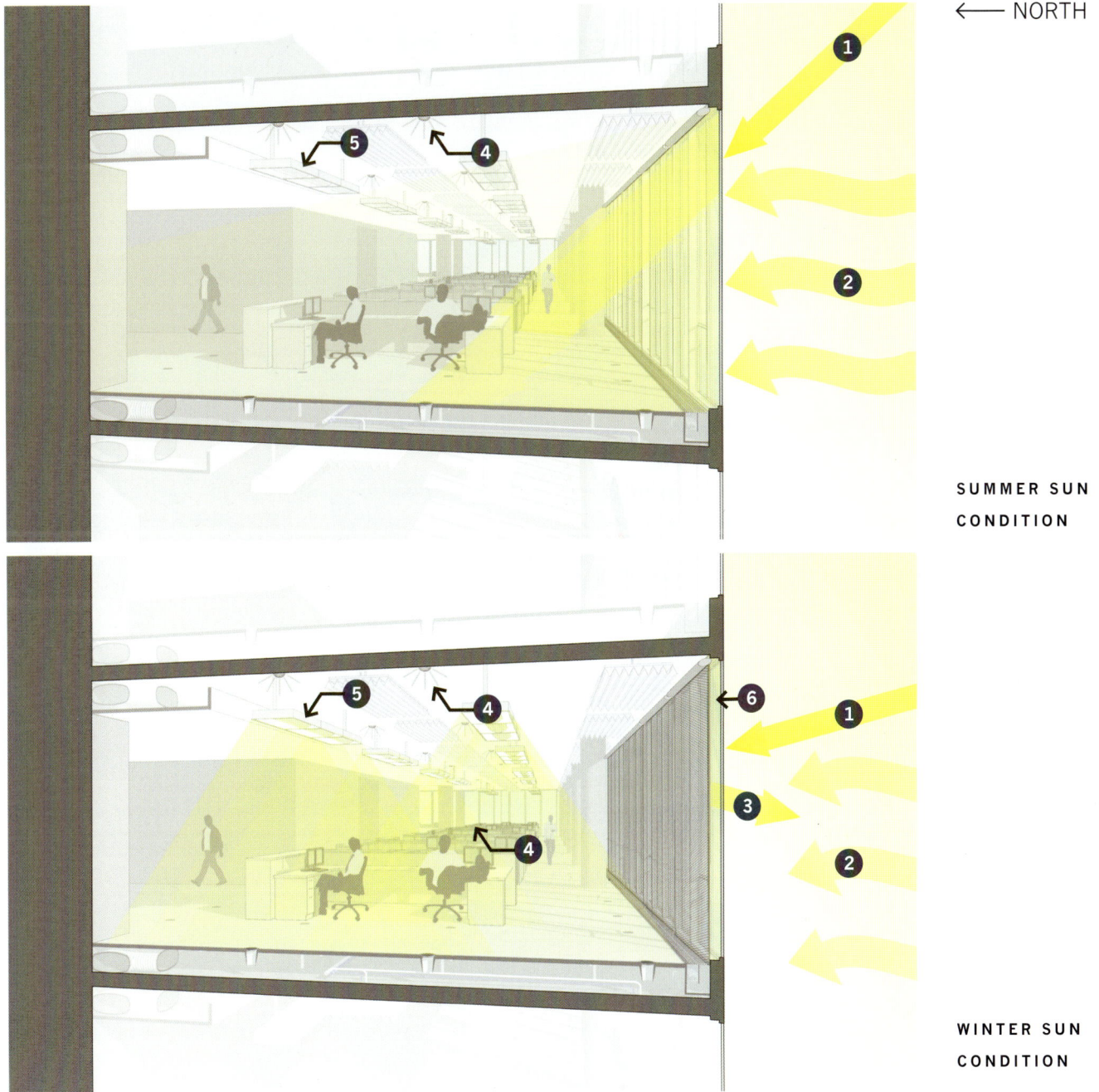

← NORTH

SUMMER SUN
CONDITION

WINTER SUN
CONDITION

HVAC: SUMMER CONDITION NATURAL VENTILATION, RADIANT COOLING

Smart design for the ZGF office floors rely on passive chilled beams for radiant cooling, which offsets solar heat gain and allows the same average temperature supply air to flow year-round.

1 Warm outside weather
2 Operable window (open)
3 Passive radiant chilled beams
4 Chilled water pipes (supply and return)
5 Individually adjustable floor diffusers

6 Under floor displacement ventilation system
7 Supply air duct
8 Return air duct
9 Fresh outside air

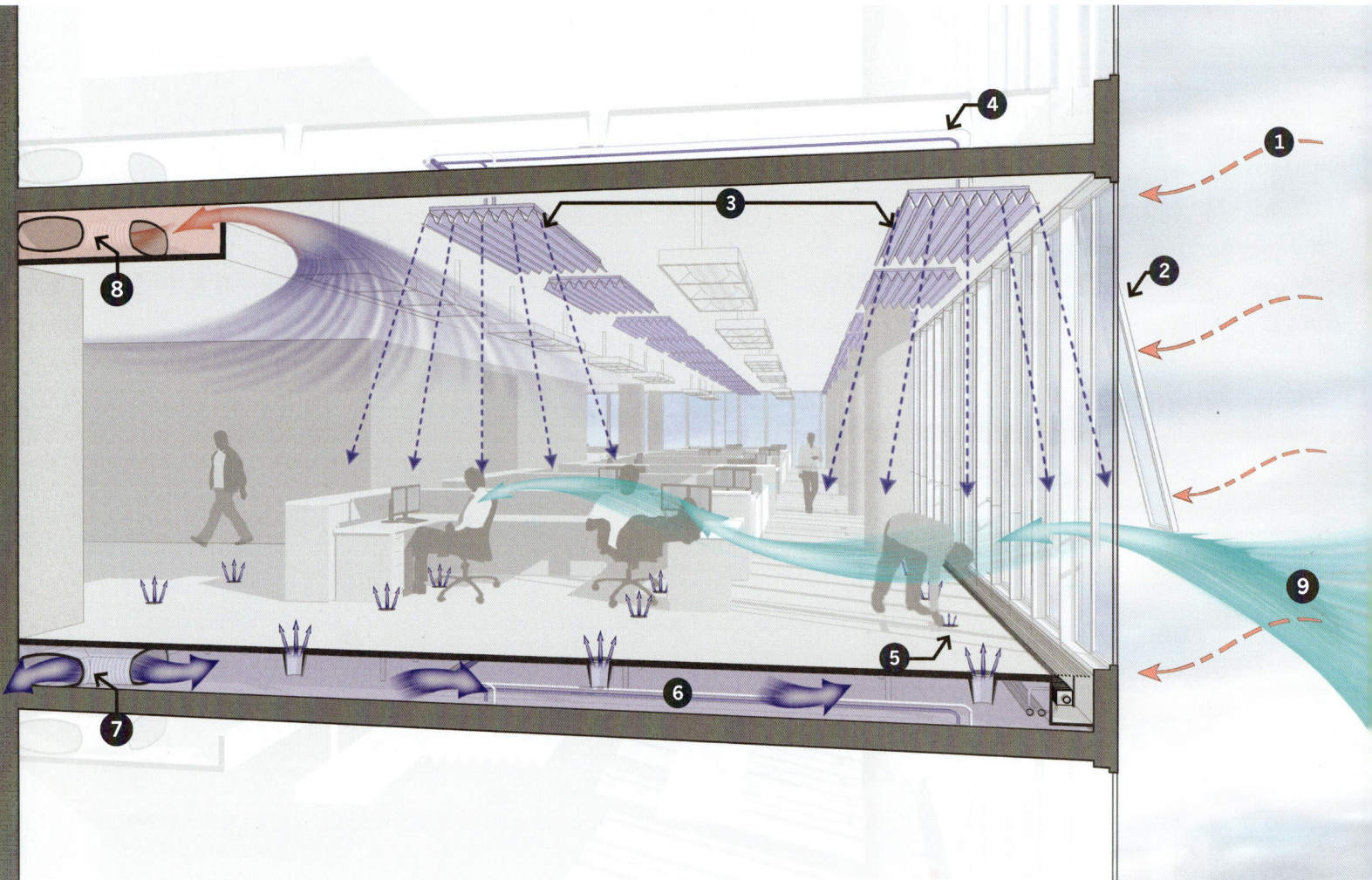

HVAC: WINTER CONDITION PERIMETER BASEBOARD HEATING

During Portland's winter months, as air contacts exterior glazing at the ceiling, it cools and drops into the baseboard's empty half – and is then pulled up through the heating element by convection and heated. This perimeter convection cycle neutralizes the cooling effect of outside weather, allowing the same average temperature supply air to flow year-round.

1 Cold outside weather
2 Operable window (closed)
3 Perimeter baseboard heater
4 Hot water pipes (supply and return)
5 Individually adjustable floor diffusers

6 Under floor displacement ventilation system
7 Supply air duct
8 Return air duct

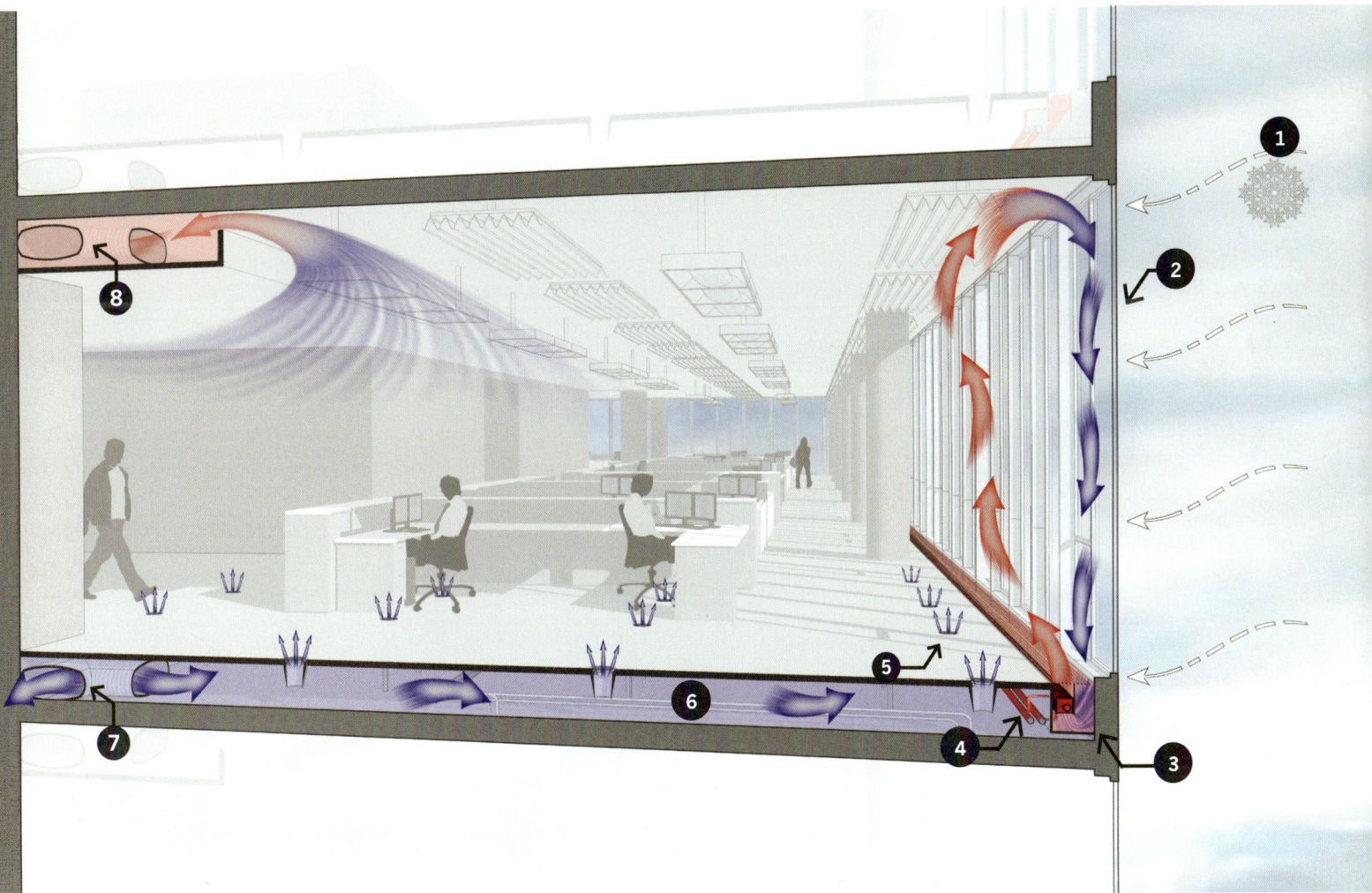

"CityCenter may very well change the way the hotel world looks at sustainability. For the first time in the lodging business at this kind of a scale, a company set out to incorporate wide-ranging green efforts in order to create some of the most environmentally friendly structures that have ever been constructed."

— Glenn Haussman

"CityCenter is Sustainability Game Changer," Hotel Interactive, December 22, 2009, www.hotelinteractive.com/article.aspx?articleid=15763

MGM MIRAGE CITYCENTER

LAS VEGAS, NEVADA

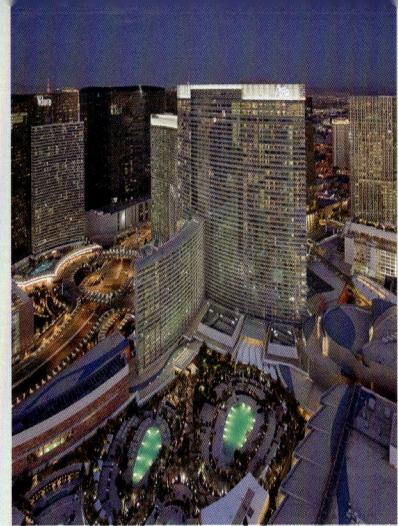

Billed as the world's largest LEED development, MGM Mirage CityCenter placed all bets on the sustainably-designed resorts, residences, luxury hotels, and retail/entertainment venues that now make up one of the main attractions on the Las Vegas Strip. This 67-acre city-within-a-city features an 8.5 mw natural gas co-generation plant, high performance exteriors, and an unprecedented array of water conservation and energy-efficiency initiatives. The project owner tasked Glumac with design review to maintain consistency in application and functionality on seven properties, totaling nearly 7.5 million-square feet (696,773,000-square meters) (two out of four blocks for the complex). Over the span of four years, this review led to enhanced commissioning services and then LEED prerequisite barrier leakage testing, calibration and commissioning for the underground parking structure's carbon monoxide system, and smoke control system testing and verification.

"If you took any one of these buildings individually, CityCenter was no different than any other project completed by Glumac. Because of the immense size – there were more than 8,000 people involved at one time, and so many different architectural, engineering and construction teams – it would have been easy to get lost without good communication. But we knew what needed to be done in terms of LEED requirements to get to the finish line."

— *Leonard Hovander, Principal, Glumac*

PROJECT DETAILS

Mandarin Oriental, Las Vegas (47-story non-gaming hotel and residences); Crystals (500,000-square foot (46,452-square meter) retail and entertainment district); Vdara Hotel & Spa (56-story, all-suite hotel); and Veer Towers (two 37-story condominium residences), two parking structures with underground parking (1-million-square feet [92,903-square meters])

Opened: December 16, 2009

Architects: Pelli Clarke Pelli, Kohn Pedersen Fox, Helmut Jahn, RV Architecture LLC led by Rafael Viñoly, Foster + Partners, Studio Daniel Libeskind, David Rockwell and Rockwell Group, and Gensler

Contractors: Perini Building Company, Tishman Construction

Developers: MGM Resorts International and Infinity World Development Corp.

All LEED Gold certified

"For our new office design at Glumac-Irvine, we wanted to 'walk the walk' instead of just moving into another building with no sustainable features. And now we're metering its energy performance: on paper, we were at .6 watts per square foot, beating the energy code by 40 percent; but under actual operating conditions, we're seeing .24 watts per square foot – beating the energy code by 70 percent, which is unbelievable. It's performing better than we could possibly imagine."

— *Brian Berg, Glumac*

GLUMAC OFFICES

PORTLAND, SAN FRANCISCO, SACRAMENTO, IRVINE, CORVALLIS

If the office of the future is smart, efficient, healthy, comfortable and technologically advanced, it also signifies an ideal place for collaboration, invention and innovation. Glumac's commitment to green buildings certainly extends to its own office spaces in Portland and San Francisco, both achieving LEED Silver certification. The LEED Platinum Sacramento office showcases a full array of sustainable MEP strategies – from thermal energy storage and underfloor air to daylighting, rainwater harvesting and effective use of external insulation to save energy by utilizing the thermal mass of concrete. Finally, the firm's newly-renovated Irvine space offers a living laboratory of high-efficiency MEP systems, demonstrating creative use of furniture-mounted task/ambient lighting and daylighting controls – and recognition as the first "Office of the Future" by the New Buildings Institute and Southern California Edison.

"The goal of the Office of the Future is to create a more responsive and responsible office environment that better serves tenant needs while reducing energy costs, enhancing property values and delivering a reduced carbon footprint."

— *New Buildings Institute,*
"Office of the Future"

PROJECT DETAILS: PORTLAND, OR

15,485-square foot (1439 square meter
office/tenant improvement

Completed February 2007

Architect: YGH

Contractor: Precision Construction

LEED Consultant: Green Building Services

LEED Silver certified

PROJECT DETAILS: SAN FRANCISCO, CA

11,500-square foot (1,068-square meter)
office/tenant improvement

Completed April 2006

Architect: Gensler

Contractor: BCCI

LEED Silver certified

PROJECT DETAILS: SACRAMENTO, CA

9,000-square foot (836-square meter) office
building/new construction

Completed August 2007

Architect: KMB Architecture

Contractor: Front Line Builders

Interior Design: ESC Design Group

LEED Platinum certified

PROJECT DETAILS: IRVINE, CA

8,700-square foot (808-square meter) office/
tenant improvement

Completed March 2010

Architect: Gensler

Contractor: Howard Building Corporation

LEED Platinum certified

HARVEY MUDD COLLEGE, 2013: As the first major new construction on campus in nearly 20 years, the Teaching and Learning Building will merge active learning and social spaces – a place where the arts and science converge.

CHAPTER NINE:

HARVEY MUDD COLLEGE TEACHING AND LEARNING BUILDING

ON CAMPUS, A BRIGHT NEW BEGINNING

"There was a sense the building ought to be dynamic and express the energy that exists on Harvey Mudd's campus. Tied to its sustainability goals as well, the design really called for creating new spaces that are much more open and transparent – and referencing this visible vitality – so you actually see what is happening inside of classrooms."

—John Meadows, Principal, Boora Architects

"O! for a Muse of fire, that would ascend / The brightest heaven of invention / A kingdom for a stage, princes to act / And monarchs to behold the swelling scene."

— William Shakespeare
Prologue, Henry V

NEW CAMPUS CENTER: At nearly 70,000-square feet (6,503-square meters) of programmable space, the facility creates a new focal point for campus – and respects the international style architecture of Edward Durell Stone while serving as a national model of sustainable design and construction.

CLAREMONT, CALIFORNIA

Open, inviting, unexpected, with ever-present views outside and to the San Gabriel Mountains in the distance – it offers the perfect setting for an evening recital of Chopin's "Polonaise in A-flat major" or a weekend performance of *Twelfth Night*. With classes in session, spaces throughout this newest building at Harvey Mudd College (HMC) focus instead on lectures devoted to nonlinear systems of differential equations, or perhaps the formulation of quantum mechanics.

Upon its completion in 2013, the Teaching and Learning Building gives the college a new centerpiece for the 55-year-old campus – offering flexible, technologically-advanced classrooms, lecture halls, faculty offices and public spaces to support a wide range of learning styles. It replaces Thomas-Garrett Hall (dedicated in 1962) with nearly 70,000-square feet (6,503-square meters) of programmable space for instruction, the arts, socializing and formal and informal gatherings, plus faculty office clusters and a new office for the president. The building effectively doubles the college's educational space as the new academic and social heart of campus.

Harvey Mudd College, consistently ranked number one or two among the nation's undergraduate engineering programs, offers a unique blend of collaborative, interdisciplinary teaching and learning. With departmental majors in biology, computer science, chemistry, engineering, mathematics and physics, the college pioneered a rigorous core curriculum – still in place today – that fully integrates the humanities, social sciences and the arts. With its student body growing exponentially, the college recently added two dorms while still faced with outdated instructional spaces: virtually all of it too small, inflexible and lacking in teaching technology. In the push to create new state-of-the-art facilities for students and faculty, HMC is now replacing all classroom spaces on campus (other than laboratories) in this modern four-story building.

The form, space and materials of HMC's Teaching and Learning Building reflect the heritage and dynamism of this unique institution. Yet, beyond simply housing the academic and creative endeavors within, the new building becomes a pedagogical tool and a showcase for green design: maximizing daylight and natural ventilation for most interior spaces and featuring advanced shading strategies, passive chilled beam and displacement air technology for cooling, an energy-efficient exterior façade, and exterior circulation to take advantage of Southern California's temperate climate.

ENVIRONMENTAL OBJECTIVES

In 2008, HMC President Maria Klawe signed on to the President's Climate Commitment[1] to reduce the global emissions of greenhouse gases, establishing criteria for the college that all new campus construction attain LEED Silver certification or higher. Previously, the college had built the first LEED-certified building (Sontag Residence Hall, 2003) in Claremont and then the LEED Silver Hoch-Shanahan Dining Common (2005). By the time construction commenced in June 2011, the new building was on track to achieve LEED Gold, with LEED Platinum in sight.

Similarly, as primarily a science, engineering and mathematics institution, HMC wanted the building to serve as a national model

1. American College & University Presidents Climate Commitment

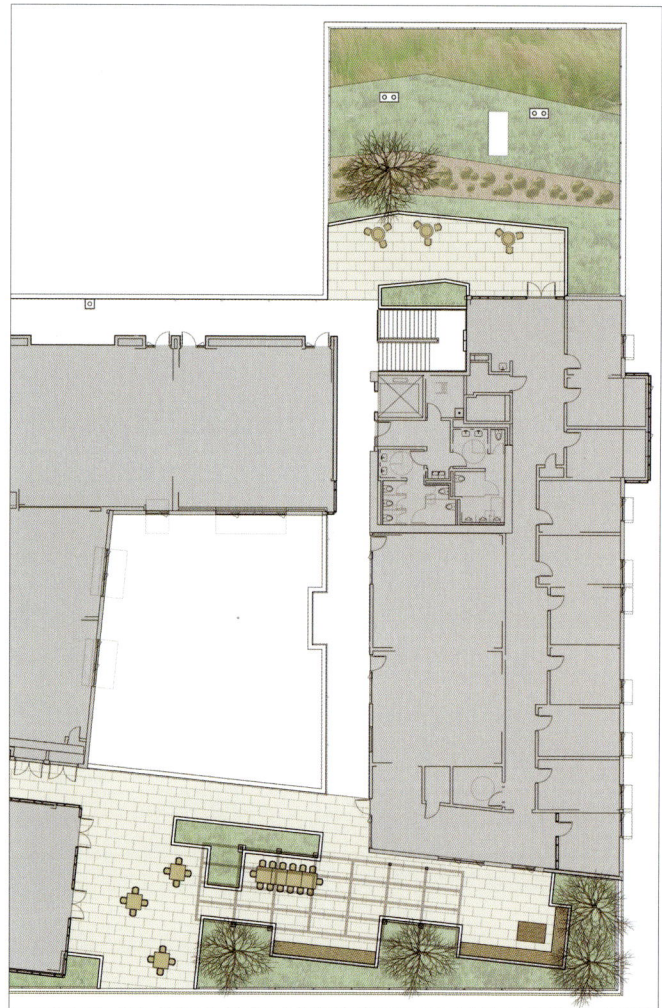

SITE CONTEXT: Adjacent to the College's
Great Mall and surrounded by heritage oak
trees, the new building assumes the same
footprint as Thomas-Garrett Hall (constructed
in 1962) – and signifies an important update
to Durell's campus master plan.

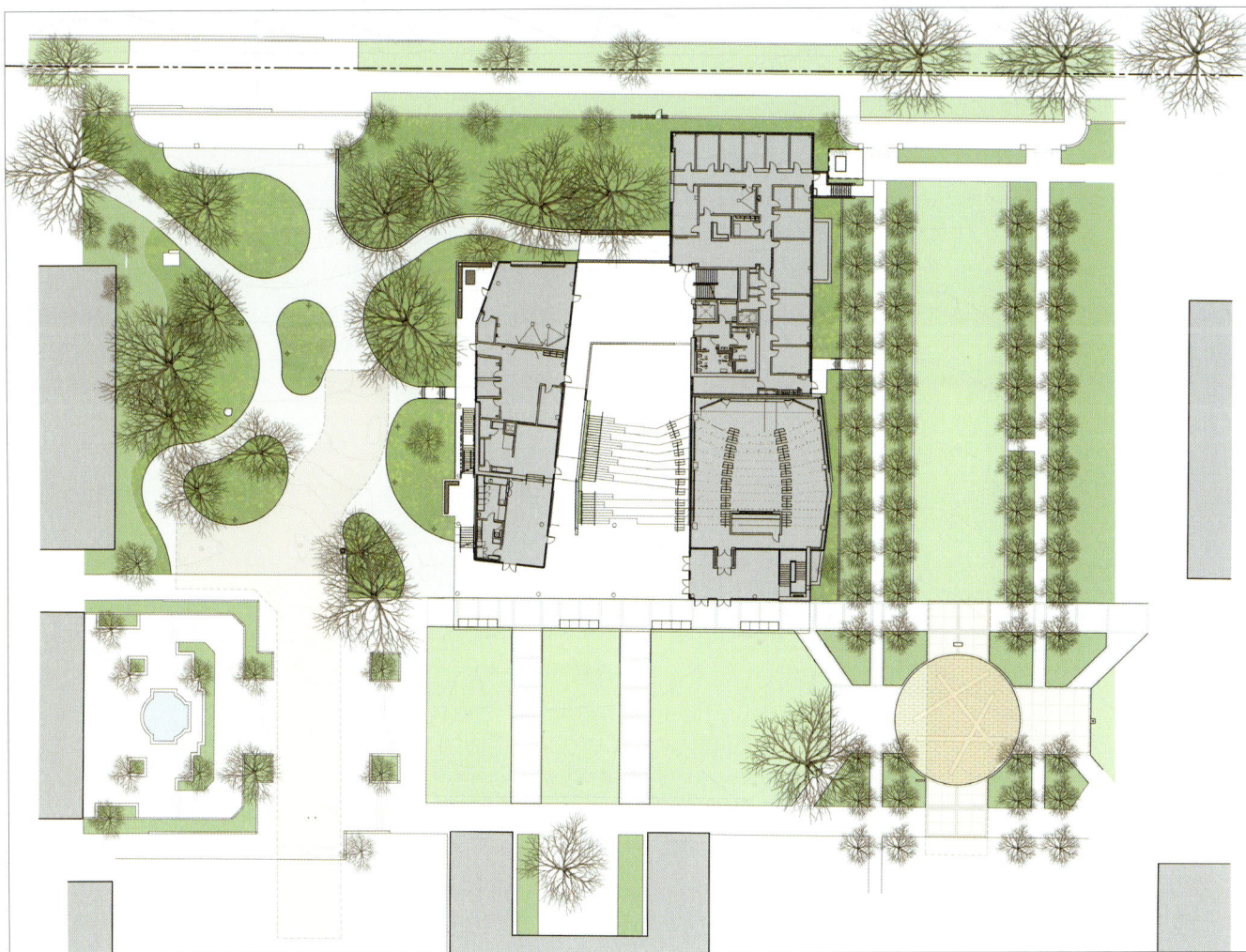

TIERED LEARNING, LISTENING: The
building's varied instructional and
performance spaces include the 100-seat
Recital Hall and 300-person Lecture Hall,
separated by a wall of glass from the adjacent
courtyard amphitheater spanning two levels.

for sustainable campus building design and construction. Passive and active design elements (daylighting, ventilation) also create "teaching moments" for students, an unprecedented opportunity to study high-performance systems – and for that performance to be visible within the classroom: to understand how structural systems work, how mechanical equipment works, etcetera.

Finally, college planners established a set of specific environmental targets for the project:

» Achieve carbon savings of 140 tons (127 kg)/year
» Reduce energy costs by 42 percent
» Reduce potable water usage by 50 percent through water-efficient landscaping
» Reduce water usage by 40 percent through efficient plumbing fixture selection
» Recycle or salvage 90 percent of construction waste
» Specify only low-emitting adhesives and sealants

DESIGN OVERVIEW

"Build for the future": the college's pledge resonates as yet another commitment to promoting its extraordinary brand of undergraduate education to the rest of the world. Two overarching themes emerged early in the design process. The first theme reflected the high priority the college places on teaching as part of efforts to recruit and retain some of the top students and faculty in the country. In particular, they wanted a facility that would demonstrate openness and transparency – especially important since most of the existing corridors and classrooms on campus are windowless.

"There are amazing things happening here obviously, and yet you could walk through the halls of academic buildings and never quite be aware of anything," noted John Meadows, AIA of Boora Architects, the lead designer. "So there was a sense that the building ought to be really dynamic and express the energy that exists on campus." Boora, based in Portland, Oregon, received the commission to design the new building after submitting a proposal in early 2009. The firm specializes in higher education projects across the country and has completed work for several colleges and universities in California, including nearby Pomona College.

The second key theme centers on honoring the existing campus architecture while looking forward to a constantly evolving future in education. Designed in the late 1950s and early 1960s by Edward Durell Stone, the entire campus exhibits three primary architectural properties: a balance of vertical and horizontal elements, repetition of the grid and concrete block, and a recurring use of pattern on exterior façades. As a result, most buildings look quite similar: rigid, yet powerful, and built out of the same materials. Stone also devised the campus master plan,

OUTDOOR CLASSROOM: Among its many unique features, the open-air rooftop terrace of the building includes an outdoor teaching space – with vine-covered pergola overhead – featuring outdoor writing surfaces.

"Given what we knew about the college, and their commitment to leadership as one of the best undergraduate engineering schools in the country, we continually asserted that Platinum was possible and within their budget – and that this higher target should really be the goal."

— *James Thomas, Vice President, Glumac*

so later buildings follow these same guidelines. Boora's design, then, evokes many of these same architectural characteristics while interpreting them in a way more connected to the interior function of the building.

The process to create the Teaching and Learning Building was collaborative and extensive. Beginning in March 2009, HMC asked Boora to lead an intensive programming study to produce a clear statement of space needs and a vision for a new building that reflects the current and future needs of faculty and students. Over the course of six weeks, this consultation included forty meetings on campus with the participation of faculty, students, trustee members, staff and administration. Sessions ranged from small groups of invited participants, targeting certain subject areas like science or math faculty, to all-campus forums held at night and open houses where anyone could attend to express their views. Additionally, the college hosted seven all-campus forums to give the larger community a chance to see progress and offer input. Boora toured instructional spaces with faculty and observed classes to better understand faculty-student interaction. Finally, the architects and project team held a one-day sustainability workshop – again inviting faculty, students, staff and trustees – as an intensive exercise to further define specific sustainable goals and opportunities.

Seven primary design goals emerged for the new building:

» Double the existing teaching and learning space
» Serve as the new focal point of campus
» Foster faculty and student interaction
» Help the college attract and retain the best faculty and students
» Nurture creativity
» Connect and unite the college community
» Serve as a national model for sustainable campus design and construction

The building's final design also influenced an important update to the campus master plan first devised by Stone more than fifty years ago. Originally, the plan showed Thomas-Garrett Hall remaining on site, utilizing a small addition to support growth. However, the administration and trustees agreed the building simply could not accommodate the college's changing academic and cultural needs. "So one of the real challenges posed to the architectural team," recalled Meadows, "became how to make a building that is as bold and exciting as the college envisions – and yet be respectful to these understated, symmetrical little buildings that would surround it."

One unifying element in the master plan is that every building features a courtyard, opening out to HMC's Great Mall.

Conceived as the new campus crossroads for academic and social activity, the Teaching and Learning Building utilizes two separate "arms" to create its large central courtyard. This flexible, open-air space is useful for many types of events and gatherings – to support the college's annual student-led Shakespearean theater and as a venue for music and art, outdoor teaching, parties or receptions, and so on.

BASEMENT LEVEL

This courtyard amphitheater tumbles down to the basement level, framed by single-loaded outdoor corridors, so that occupants move throughout the building via outdoor walkways that connect all program elements. Spaces on this level include state-of-the-art digital media and electronic music studios, a 100-seat recital hall, flexible gallery space for student and faculty artistic exhibitions, as well as a number of the building's larger classrooms, including a lecture hall and an 85-person tiered classroom.

GROUND LEVEL

One floor up, the ground level features student-oriented spaces such as the café, the student living room and Writing Center, as well as public-oriented spaces such as the Office of Admission and Financial Aid. On the southeast corner, a 300-person auditorium "steps down" with the adjacent courtyard space, separated only by a wall of glass. The auditorium spans both the basement and ground levels, creating a two-story lobby able to support larger events.

SECOND LEVEL

The next floor includes a mix of classrooms, breakout spaces and interdisciplinary faculty office clusters – carefully designed to promote interaction among students, staff, instructors and all members of the college community. The Office of the President sits one level above the Office of Admission and Financial Aid in the northeast corner.

THIRD LEVEL

The building's top floor includes additional classrooms and faculty offices, as well as two distinct open-air terraces. Designers planned the north terrace as a quiet, landscaped, contemplative space, with views to the San Gabriel Mountains. In contrast, they created the south terrace as an active space, featuring a vine-covered pergola and an outdoor teaching space with outdoor writing surfaces. At the southwest corner sits the "prow," a large 50-person classroom clad in glass and offering immediate access to the open-air terraces – an ideal space as a dual-purpose room for events and receptions.

MEP DESIGN PROCESS

Boora credits the project's integrated design process for arriving at many fundamental decisions on structural and MEP systems.

CROSS-DISCIPLINARY: Four-story building section illustrates flexible, technologically-advanced classrooms, lecture halls, faculty offices and public spaces to support a wide range of learning styles – effectively doubling the College's educational space.

With input from Glumac as MEP designer and KPFF Consulting Engineers as structural engineer, the architect could explore passive solutions that were "inherently energy efficient and delightful," said Meadows – solutions which then influenced the building layout, orientation, window positioning, and more. This design results in a very thin building – a U-shape along the east, west and south sides that wraps around the courtyard. In turn, many of the indoor spaces feature daylighting, often on two sides, while rooms throughout the building can maximize the use of operable windows for fresh air and cross ventilation. From the outset, the team wanted to achieve the high performance of LEED Platinum at the same budget as a much more conventional building. "So even before adding chilled beams, controls and other equipment," he stated, "we could be far enough in terms of energy savings to afford the technology."

Another fundamental precept for the building emphasizes its use as a learning tool, from design through construction and operation. As a new space for Harvey Mudd engineering students, the college wanted to tap their energy and ideas during the design charrettes and early in the SD (schematic design) phase of the project.

Boora also hired three HMC students as research interns for summer 2009 to work with designers in Glumac's Los Angeles and Portland offices on a variety of sustainability concepts.

David Summers, a Glumac Principal in Los Angeles and a Harvey Mudd graduate, collaborated with the students to discuss system concepts and brainstorm innovative solutions. In addition to daylight modeling, the students conducted general research on campus water conservation and photovoltaic studies. They also used COMFEN, new schematic design software developed by Lawrence Berkeley Labs, to perform quick analyses of façade and shading scenarios to gauge the building's energy performance and heating and cooling requirements.

The new Teaching and Learning Building also incorporates smart meters that feed data into the building management system to record chilled water usage and lighting, mechanical and plug loads, along with temperature readouts from the air handling unit. Much of this information, in turn, will transmit to a building dashboard/user interface, to be created by HMC students. While at the Building, occupants who sign in to use the Internet will be directed to the dashboard automatically as a default homepage. Additionally, a prominent display in the Building's café will stream real-time building performance on classroom space conditions, energy loads, lighting loads and chilled water capacity for any given month – all available as downloadable reports, spreadsheets and modeling data for use in classroom projects. The design team and campus administrators hope that visualizing this data will increase awareness among occupants, encouraging them to turn off lights or open and close windows, further improving performance.

INDOOR AIR

Among the primary air strategies, Glumac engineers chose displacement ventilation for classrooms and the large lecture hall (3,500-square feet [325-square meters]) on the basement level. These systems distribute cool air at low velocities within both the tiered (through raised access floors) and non-tiered (through specialized corner-mounted grilles) classrooms. By supplying air at the floor level and conditioning only occupied zones, designers cut the amount of air needed in half – approximately 5,000 cfm – whereas a conventional VAV system would require as much as 10,000 cfm. The building's displacement system also allows for higher supply air temperatures, from 64 to 75°F (17.78 to 23.89°C, providing many more hours of free cooling in California's climate.

A single air handler serves the entire building with 100 percent outside air. Designers sized the system to produce excess outside air/ventilation air by 30 percent or greater, consistent with ASHRAE 62, the indoor air quality standard.

Initially, Glumac recommended displacement ventilation throughout the building – a scheme that realizes additional energy savings when combined with chilled beams; yet the added cost of displacement diffusers required the engineers to later scale back the use of displacement ventilation as part of value engineering measures. The ground through third levels

MIXED MODE COOLING

COOLING STRATEGIES: Glumac's integrated design approach focuses on both passive and active strategies – which influenced building layout, orientation and fenestration – through use of operable windows for cross ventilation, displacement ventilation, chilled beams, and manual HVAC controls.

1   Dedicated outside air duct
2   Supply diffusers
3   Passive chilled beam
4   Operable windows
5   Supply air plenum
6   Vertical displacement diffusers
7   Return duct
8   Wall displacement diffusers

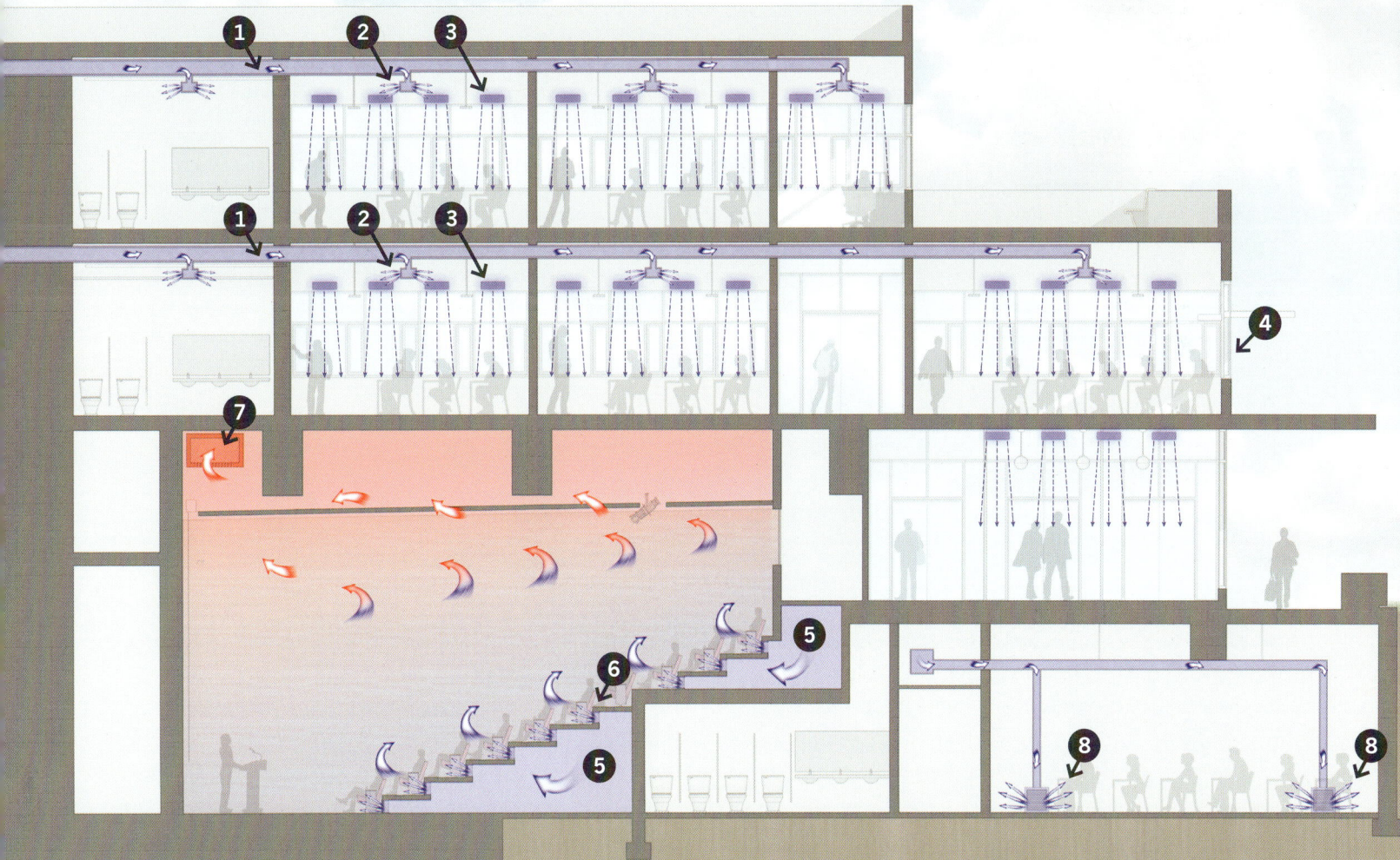

VARIATIONS ON A THEME

COMFORT CONTROL: The College and project team gave careful consideration to designing the Teaching and Learning Building as a pedagogical tool and showcase for sustainability – in turn, incorporating manual procedures and automated controls responsive to outside conditions and seasonal variations.

1 Dedicated ventilation air duct
2 Ceiling fan
3 Passive chilled beam
4 Occupant controls for mechanical cooling and fans
5 Operable windows

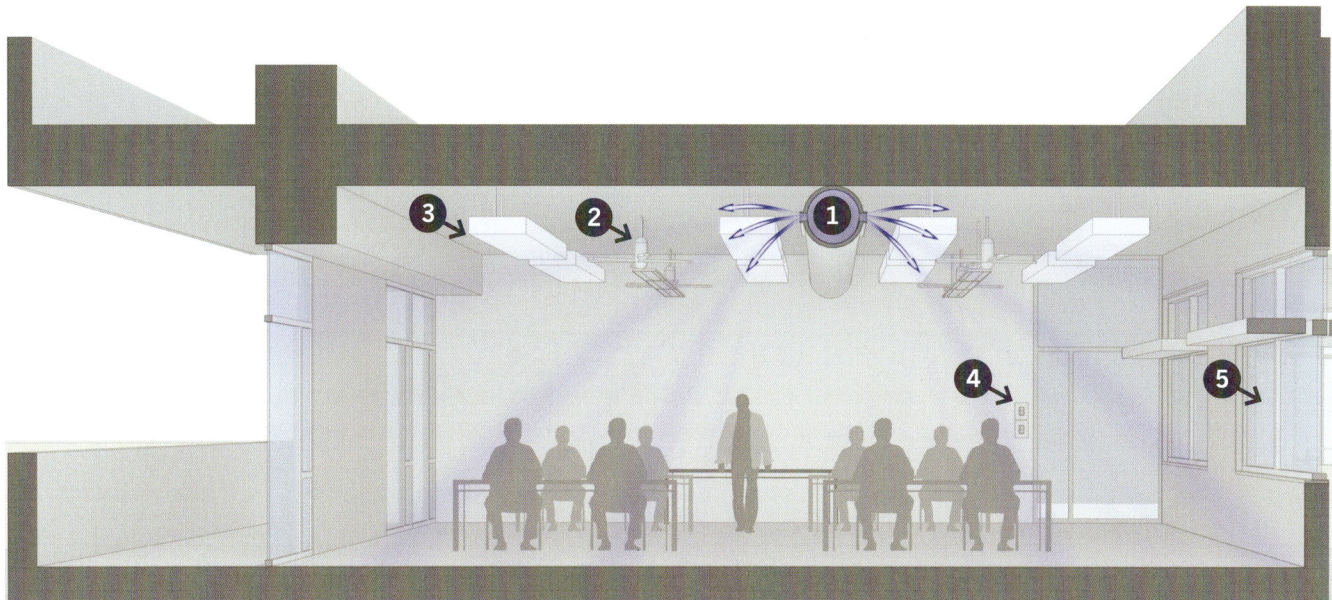

ECONOMIZED ENVIRONMENT: With comfortable outside air temperatures, ventilation air may be supplied as is, without further conditioning.

• Outside air temperature 60°F
• Chilled beams off
• Dedicated outside air economized and exceeds ventilation requirements

OCCUPANT CONTROLLED ENVIRONMENT: With comfortable outside air temperatures, occupants may use manual controls to deactivate mechanical cooling systems and turn on fans while opening windows for natural ventilation. The mechanical system reverts to a default setting one hour after activating the switch.

- Outside air temperature 70°F
- Chilled beams off
- Dedicated outside air off and windows open

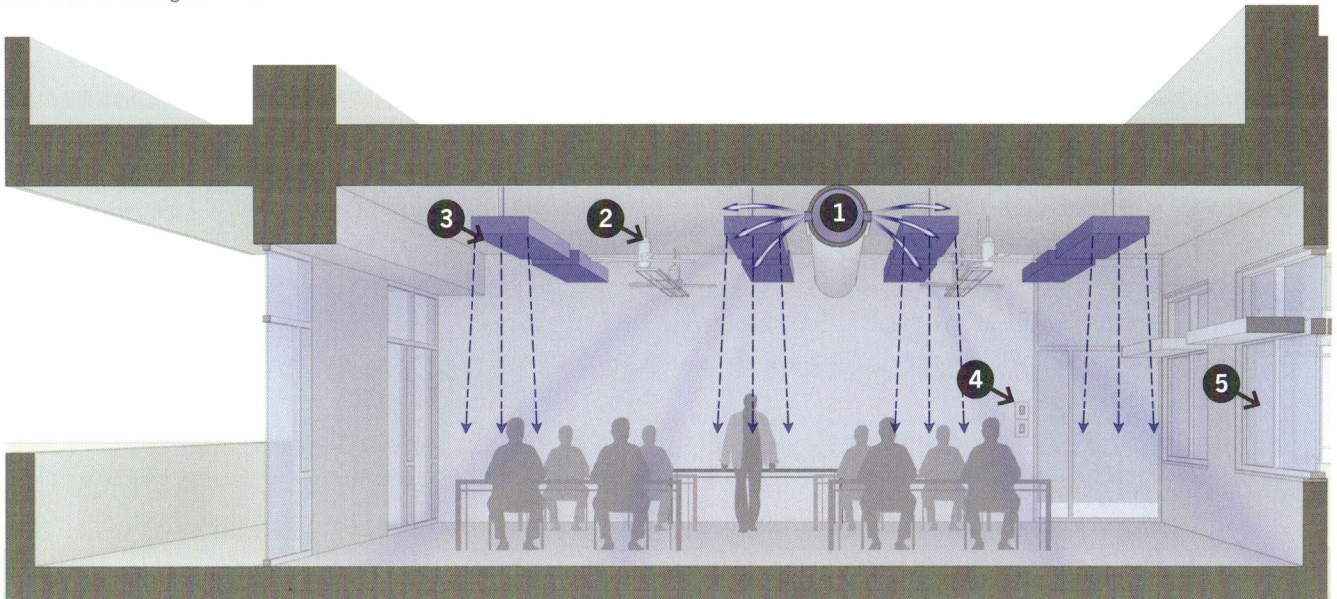

MECHANICALLY COOLED ENVIRONMENT: In the event of uncomfortably warm outside air temperatures, the system activates chilled beams to provide cooling. For additional cooling as needed, dedicated outside air may be chilled before entering the space.

COOLING PHASE ONE
- Outside air temperature 90°F
- Chilled beams on
- Dedicated outside air meets ventilation requirments

COOLING PHASE TWO
- Outside air temperature 90°F
- Chilled beams on
- Dedicated outside air chilled for auxillary cooling

"This courtyard creates a magical moment – and a bit of a surprise. As you enter under this big porch and come into the courtyard, out emerges this wonderful space, this line to the classroom, with nooks for students, coffee shops, and all sorts of activities."

— *John Meadows, Boora Architects*

therefore feature operable windows in classrooms, offices and tiered lecture spaces to support natural ventilation. These rooms also contain conventional overhead diffusers to help promote air movement and meet comfort requirements; several offices with limited ventilation air also include high induction diffusers to increase air movement.

Designers also explored operational issues for integrating the windows with HVAC systems to control conditions within the building's wide array of room types, sizes and functions. With the college, they considered a fully automated system using window sensors to shut off heating or cooling, an entirely manual system, and a green light/red light approach to alert users when to open or close windows. Ultimately, the college elected to include manual wall-mounted switches on time-delays, adjustable to fit class schedules; this choice reinforces their efforts to educate building users on how systems work and their ability to control comfort, air quality and energy savings within the building.

WATER & WASTE

Glumac engineers selected plumbing fixtures based on reliability, durability and low-flow features. Water-efficient systems throughout the building include: wall-mounted water closets, 1.28 gpf; urinals, 0.125 gpf; lavatories, 0.5 gpm; and showers, 1.5 gpm.

HEATING & COOLING

With the site's existing building already served by the campus chilled water plant, the college chose to retain that service connection for the new Teaching and Learning Building. The basement level displacement ventilation systems supplies cooling or heating accordingly via a chilled water coil or heating coil in the building's air handler.

Passive chilled beams will supply cooling for the remaining offices and classrooms as well as the Writing Center and lounge. Glumac had considered radiant panels and chilled sails, finally settling on chilled beams with a linear face as the most efficient, cost-effective technology per BTU of cooling. Demand control using CO_2 measurement in each classroom minimizes energy use from these zones.

A study conducted by HMC students concluded the new building would not require *any* heat. Their rationale: occupancy in classrooms and other spaces would drive up temperatures on most days, while the number of occupied hours during the school year where temperatures fell below 65°F (18.33°C) (the comfort threshold) remained very low. However, Glumac's engineers still recommended adding heat for occupant comfort and specified European-style hot water radiators along the perimeter. Budget restrictions later eliminated this option, so that heat will be supplied by the heating coils within the ventilation terminals.

MAXIMUM DAYLIGHTING: The open design, driven largely by the building's large central courtyard, made it possible to daylight more than 50 percent of spaces – a scheme using COMFEN software to perform quick analyses of façade and shading scenarios to gauge energy performance.

LIGHTING & DAYLIGHTING

Over 50 percent of the Teaching and Learning Building is daylit – including all classroom spaces on the top floors. Even most basement-level classrooms feature either a window into the courtyard (Recital Hall, Lecture Hall) or a skylight (classrooms on the south). Glumac performed all lighting studies to determine daylighting potential and all calculations for the site, per California's Title 24, to address window area, daylight penetration, skylight placement and other parameters. Glumac engineers complemented this system with a series of automatic lighting controls, programmed to seamlessly dim fixtures within daylit spaces in response to available outside light. Additional measures taken to control daylight include external shades on the east side façade outside every office and shading elements on the south side classrooms.

Glumac also collaborated with an outside lighting designer on interior and exterior systems and controls. To balance out the daylighting scheme, the designer specified high-efficiency luminaires and suggested the architect maximize spacing and limit the number of fixtures to minimize energy use, in addition to occupancy sensors and daylight photocells throughout. Glumac also coordinated the location of pendants and recessed luminaires with chilled beams and ceiling fans to avoid spatial conflicts, undue shadows or reflected glare effects.

SITE

The building design preserves and highlights several mature oak trees on the north and west edges of the site. Landscaping surrounding the building and on roof terraces relies predominately on drought-tolerant, California native plants.

ENERGY

Lighting, receptacle and HVAC electrical loads will be metered to measure overall building energy usage and monitor the efficiency of operations for troubleshooting purposes. In turn, all data that feeds into the building dashboard can be downloaded from the building automation system.

As a significant future addition, the college allocated 4,100-square feet (381-square meters) of space on the roof for photovoltaic panels, expected to offset approximately 12 percent of the building's energy needs. Electricity generated by the PV array will be sub-metered for informational/educational purposes. Also under consideration, HMC may supplement its on-site power generation with a future geoexchange system.

"There's a lot of blue sky thinking happening with the South Lake Union redevelopment – so that we were encouraged to investigate many sustainable concepts, then bring them down to commercial reality in considering whether they could be applied or not. It's all new technology, and very, very few people have experience with these kinds of projects. Consequently, we're still very much at the steep end of the learning curve."

— *Marc Jacques, Associate Principal, Glumac*

2201 WESTLAKE

SEATTLE, WASHINGTON

Seattle's South Lake Union neighborhood has certainly attracted its share of signature green buildings in recent years. Add to that growing list the 2201 Westlake development, featuring 300,000-square feet (27,900-square meters) of office space, 135 luxury condos known as "Enso", 25,000-square feet (2,322.58-square meters) of street-level retail, and underground parking. Integral to the revival of this downtown lakefront neighborhood, one of the city's oldest, the building also represents the first mixed-use, high-rise residential project to earn LEED Gold certification in Seattle. Throughout, 2201 Westlake emphasizes daylighting, low VOC finishes, water-efficient fixtures, and recycled and rapidly renewable materials. All office levels include underfloor air delivery for heating, cooling and ventilation. And perhaps its most innovative design element: a shared, high-performance central energy plant, which moves waste heat and cooling between the office space and adjacent condominium tower according to the time of day.

"...the 'hood' is home to a trove of new businesses along Westlake Avenue and a fresh assemblage of urban explorers, thanks in part to Vulcan Real Estate's new 2201 Westlake building, which houses offices of Amazon.com and global health giant PATH."

— *Eran Afner, "Urban Safari: Denny Triangle," Seattle Magazine, August 2010*

PROJECT DETAILS

Seven-story, 400,000-square foot (37,200-square meter) office, thirteen-story condominium

Completed July 2009

Architect: Callison

Contractor: Sellen Construction

Owner: Vulcan Northwest

LEED Gold certified

"These are the traits of sustainable businesses: holistic enterprises that regard the health of their community, environment, and economy as being of fundamental and equal importance. This particular property was designed with these ideals top of mind..."

— *Liz Mazurski*
"Growing Green at The Allison Inn & Spa in Oregon," Spa Magazine
www.spamagazine.com/articles/oregon/growing-green-allison-inn-spa-oregon

THE ALLISON INN & SPA

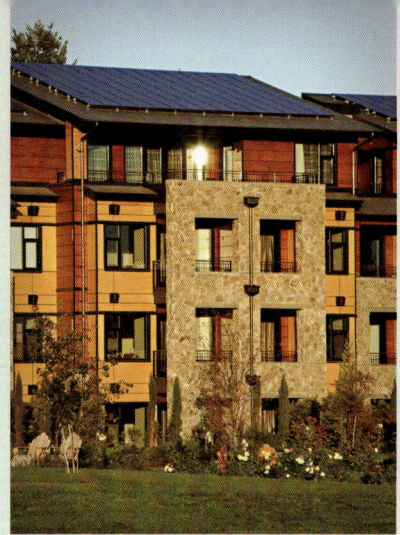

Oregon's Willamette Valley wine region is known for its rich red soils, famed pinot noir and, increasingly, sustainably-designed destinations like The Allison Inn & Spa. Located just twenty-five miles from Portland, this 85-room luxury resort also features a signature restaurant, 12-treatment-room spa, covered indoor pool, and 14,000-square feet (1,300-square meters) of meeting and social gathering space. Envisioning an energy-efficient model for the hospitality industry, owners Ken and Joan Austin agreed to aggressive goals for water, energy and materials performance on the project. The resulting suite of sustainable measures anticipates 40 percent less water use and 60 percent less energy consumption compared to a baseline building under ASHRAE 90.1-2004. Glumac's designs include highly efficient plumbing, solar water heating and dedicated outside air systems, plus energy consulting on a series of rooftop photovoltaic panels to generate 7 percent of the Inn's electricity needs.

"The owners' commitment to go green was very driven and directed towards achieving LEED Gold, and they were always open to a lot of suggestions. Everyone on the project was willing to push, to see what could be accomplished. In the end, we achieved a highly sustainable design approach – and we left few stones unturned."

— *Eliot Jordan, Glumac*

PROJECT DETAILS

Four stories (east wing) and two stories (west wing); 155,000-square feet (14,400-square meters)

Completed September 2009

Architect: GGLO

Contractor: Lease Crutcher Lewis

Owner/developer: Springbrook Properties, Inc.

LEED Gold certified

"Historically, the data center industry has not been able to accept the idea that reliability *and* energy efficiency can co-exist in the same facility. But that perception is starting to turn around – and here, Glumac has been able to show an enterprise client how to integrate sustainable solutions for data center design."

— *Sam Graves, Associate Principal, Glumac*

SUN DATA CENTER

SANTA CLARA, CALIFORNIA

A model high-density, next-generation data center: that was the chief objective of Sun Microsystems (now part of Oracle) in consolidating facilities on its Santa Clara campus. To meet tight space and budget targets, the server manufacturer identified equally aggressive energy efficiency and environmental goals for the new facility. With breakthrough mechanical designs led by Glumac, the data center features air-side economization (using outside air) to deliver free cooling 75 percent of the time, while high chilled water temperatures enable chillers to run at greater efficiencies. The building's high-density layout (850 watts per square-foot or 9.1 kilowatts per square-meter) also demonstrates effective use of "hot aisle containment," pulling high temperature air away from server corridors without circulating back into the space. Overall, Sun estimates annual electricity costs to fall by $1.1 million as a direct result of enhancements in power efficiency.

"By focusing on the biggest operating expenses – the data center – Sun has been able to deliver cost-effective solutions for CIOs facing a declining budget with increasing demands for new services and products."

— *Sandra Kay Miller, "Cost Savings In The Data Center,"* Processor, *May 30, 2008, page 27*

PROJECT DETAILS

55,000-square foot (5,100-square meter) high-density data center (with Tier II requirements)

Completed December 2007

Architect: Gensler

Contractor: Devcon

THE APPENDICES

PROJECT INFORMATION

GERDING THEATER AT THE ARMORY

Portland, Oregon

PROJECT PROGRAM & STATS

Building Type: Assembly/Theater

Size: 55,000 square feet (5,110 square meters)

Scope: Lobby, 600-seat main stage theater, 200-seat "black box" theater, rehearsal halls, community space, Portland Center Stage office space, gallery and café

Stories: Three

Sustainable Design Features:

Displacement ventilation: underfloor air distribution systems in the lobby, theaters and administrative offices, plus "active" chilled beams placed strategically on several floors supplement both cooling and heating

» Forty-two skylights, seventeen of them operable, admit natural light and fresh air into offices, rehearsal halls and the entrance lobby; advanced glazing maximizes daylighting while minimizing winter heat loss and summer heat gain

» Dimmable compact and linear fluorescent fixtures provide general illumination, with lighting controls that rely on "open-loop" photoelectric daylight sensors and "closed-loop"/occupancy sensors

» A rainwater catchment system diverts gray water to flush toilets and urinals; overflow from the system's 12,000-gallon (45,425-liter) storage tank and a portion of the sidewalk runoff drains to irrigate native plantings

» Existing mass of the restored 1891 historic building, along with new concrete floors and walls, serves as a thermal "flywheel" to reduce diurnal temperature swings

Completion Date: September 2006

Cost: $28 million

Ratings: LEED® Platinum (LEED-NC, v.2/v.2.1)

PROJECT TEAM

Client/Owner/Developer: Portland Historic Rehabilitation Fund

Developer: Gerding Edlen

Glumac Design Team: MEP Engineering, Energy Consulting, Commissioning, Lighting Design

Architect and Interior Designer: GBD Architects, Inc.

Design Consultants: *Environmental Building Consulting*: Green Building Services, Inc.; *Civil and Structural Engineering*: KPFF, Inc.; *Landscape Architecture*: Murase Associates; *Theater Consulting*: Landry & Bogan, Inc.

General Contractor: Hoffman Construction Company

SELECTED AWARDS

2008

Preservation in Action Pinnacle Award, Architectural Heritage Center/Bosco-Milligan Foundation

2007

AIA/COTE Top Ten Green Projects, Honorable Mention, American Institute of Architects (AIA)/Committee on the Environment (COTE)

Urban Land Institute, Awards for Excellence: The Americas

"America's Greenest Buildings," *Forbes Magazine*

KELLEY ENGINEERING CENTER

Corvallis, Oregon (Oregon State University)

PROJECT PROGRAM & STATS

Building Type: Academic/Research

Size: 153,000 square feet (14,214 square meters)

Scope: 155 faculty and graduate student offices, 2,200 open
computer spaces, 12 conference rooms, two large theater-
style classrooms, two "reconfigurable" classrooms/conference
rooms, nine seminar classrooms and central atrium

Stories: Four

Sustainable Design Features:

» The central atrium is designed for daylighting and natural
ventilation, with air flowing from perimeter offices and GRA
spaces, utilizing the 74-foot-high (22.56 meter) space (acting
as a stack-driven chimney), then out through external exhaust
louvers; the atrium also features low intake openings; night
flushing from motorized openings pushes a large volume of
air through the building by passive means, resulting in an
overall fresher building in the morning while delaying the
start of mechanical cooling

» An underfloor air distribution system for offices, labs and
work spaces delivers a constant temperature and aids with
indoor air quality, cooling, heating and energy savings

» Daylight illuminates most classrooms, labs and offices
adjacent to the atrium; south-facing exterior sunscreens
minimize heat gain; interior light shelves control glare

» A rainwater harvesting system is used for toilet and urinals
and on-site irrigation, while water-efficient fixtures reduce the
building's water usage

» Building automation includes occupancy sensors for climate
and lighting control; perimeter offices feature daylight/
dimming controls and motorized operable windows

» The rooftop 2,400-watt photovoltaic array offers a solar
demonstration and screen wall for mechanical equipment,
while an evacuated tube solar hot water collector system
supplies domestic hot water

Completion Date: 2004

Cost: $45 million

Ratings: LEED Gold (LEED-NC, v.2)

PROJECT TEAM

Client/Owner: Oregon State University

Glumac Design Team: Mechanical and Electrical Engineering,
Lighting

Architect: Yost Grube Hall Architecture

Design Consultants: *Program and Lab Consulting*:
SmithGroup Inc.; *Structural and Civil Engineering*:
KPFF Consulting Engineers; *Landscape Architecture*:
GreenWorks; *Sustainability Consulting*: Green Building
Services; *Audio Visual/IT/Security*: Netsystems; *Signage
Consulting*: The Felt Hat; *Cost Estimating*: C3 Management
Group; *Acoustical Consulting*: Altermatt Associates,

Inc.; *Smoke Control/Life Safety*: Rolf Jensen Associates; *Commissioning*: CH2M Hill

General Contractor: Skanska USA Building, Inc.

SELECTED AWARDS

2006

Outstanding Ceiling Project of the Year Award, Northwest Wall and Ceiling Bureau

2005

Sustainable Merit Award, AIA/Portland Design Awards

Honor Award for Educational Interiors, IIDA Portland Chapter

2003

Gold Award, Sustainable/green, Portland Design Festival

WAYNE L. MORSE UNITED STATES COURTHOUSE

Eugene, Oregon

PROJECT PROGRAM & STATS

Building Type: Government/Judicial Complex

Size: 270,000 square feet (24,084 square meters) (building), 4.2 acres (16,997 square meters) (lot)

Scope: Courthouse (first and second floor offices, interior courtyard, third floor courtrooms, fourth floor judges' chambers, judicial library spaces), underground parking garage, computer rooms

Stories: Five above grade, One below grade

Sustainable Design Features:

» The low-velocity underfloor air distribution (UFAD) system supplies a majority of spaces in the courthouse – in addition to "air walls," which produce supplemental displacement ventilation and cooling for public areas

» Coupled with the building's displacement ventilation scheme, a radiant-floor system in lobbies and public spaces supports 100 percent of the heating load in winter and provides partial cooling in summer

» Multiple air handlers, configured as fan walls, increase operating efficiency and reduce noise as an alternative to large house centrifugal fans

» Large exterior windows and high ceilings within the building encourage daylighting – including within each of its six courtrooms
» Exterior sunscreens, high performance glazing and shading structures manage solar heat gain
» Waterless and low-flow fixtures, combined with native and drought-tolerant plants for landscaping, reduce total water use by more than 40 percent

Completion Date: November 2006

Cost: $78 million

Ratings: LEED Gold (LEED-NC, v.2/v.2.1), Energy Star

PROJECT TEAM

Client/Owner/Developer: U.S. General Services Administration

Glumac Design Team: Mechanical/Plumbing Engineering, Commissioning

Architects: DLR Group (Executive), Morphosis (Design)

Design Consultants: *Sustainability Advisor/LEED Consulting*: Brightworks; *Civil and Structural Engineering*: KPFF, Inc.; *Electrical Engineering*: DLR Group; *Landscape Architecture*: Richard Haag Associates, Inc.; *Lighting Design*: Horton Lees Brogden

General Contractor: JE Dunn Construction

SELECTED AWARDS

2009

AIA Northwest & Pacific Region Design Awards, Honor Category

2008

Design Excellence Awards, Honor Awards: "Architecture," "Construction Excellence," and "Art in Architecture," General Services Administration

Platinum Award, Building Team Awards, *Building Design + Construction Magazine*

2007

AIA/COTE Top Ten Green Projects, American Institute of Architects (AIA)/Committee on the Environment (COTE)

Citation, Environmental Award for Green Strategies, General Services Administration

Citation, AIA Academy of Architecture for Justice (AAJ) · *Justice Facility Review*

Distinguished Building Award, Chicago Athenaeum Museum of Architecture · American Architecture Awards

2005

BE Awards of Excellence, "Building: BIM for Architecture, Public Building," Bentley Systems, Incorporated

TWELVE | WEST

Portland, Oregon

PROJECT PROGRAM & STATS

Building Type: Mixed-Use, High-Rise Office/Housing/Retail

Size: 552,000 square feet (51,282 square meters)

Scope: Street-level retail space, four floors of office space leased to ZGF Architects, seventeen floors of apartments and penthouse homes, and five levels of below-grade parking along with three roof-level terraces and gardens

Stories: Twenty-two above grade, Five below grade

Sustainable Design Features:

» Daylighting strategies for 90 percent of occupied spaces rely on low-e vision glass and a centralized control system for occupancy, automated sunshades and demand response to adjust lighting levels and manage solar gain

» The building's concrete thermal mass, natural ventilation through operable windows, and an underfloor air distribution system combine to provide cooling for the four office levels; passive chilled beams produce additional perimeter cooling within ZGF's occupied spaces

» Nightflushing, plus fanwall technology/economizer-mode air handlers used to deliver up to 100 percent outside air, enhances cooling, indoor air quality and energy savings

» A rainwater harvesting system and low-flow fixtures manage storm water, meet 100 percent of the green roof's irrigation needs, and reduce potable water use by more than 44 percent

» Four building-mounted wind turbines on the rooftop generate one percent of electricity requirements – enough to power the building core – while solar thermal panels provide 24 percent of the building's hot water needs

Completion Date: July 2009

Cost: $138 million

Ratings: LEED Platinum (LEED-NC, v.2 and LEED-CI v.2)

PROJECT TEAM

Client/Owner: Gerding Edlen

Glumac Design Team: MEP Engineering, Lighting Design

Architect, interior Design, Landscape Architecture: Zimmer Gunsul Frasca Architects LLP (ZGF)

Design Consultants: *Structural Engineering*: KPFF Consulting Engineers; *Civil Engineering*: David Evans and Associates; *Acoustics*: Altermatt Associates; *Curtain Wall:* Benson Industries; *Wind Turbines*: Southwest Windpower; *Solar Hot Water Panels*: Heliodyne Inc.

General Contractor: Hoffman Construction Company

Developers: Gerding Edlen. Downtown Development Inc.

SELECTED AWARDS
2010

AIA/COTE Top Ten Green Projects, American Institute of Architects (AIA)/Committee on the Environment (COTE)

Built Merit Award, AIA Portland Design Awards

Best in Show, TopProjects, *Daily Journal of Commerce*

2009

Smart Environments Awards, International Interior Design
Association/*Metropolis Magazine*

TEACHING & LEARNING BUILDING, HARVEY MUDD COLLEGE

Claremont, California

PROJECT PROGRAM & STATS

Building Type: Academic/Multi-Use

Size: Approximately 70,000 square feet (6,503 square meters)

Scope: Courtyard – framed by single-loaded outdoor corridors, digital media and electronic music studios, 100-seat recital hall, gallery space, lecture halls, café, "student living room", Writing Center, Office of Admission and Financial Aid, 300-person auditorium, tiered and non-tiered classrooms, faculty office clusters, Office of the President, rooftop terraces, outdoor classroom

Stories: Three above grade, One below grade

Sustainable Design Features:

» Building layout, orientation and fenestration allows for cross-ventilation and maximizes daylighting (often on two sides)

» Basement level displacement ventilation delivers low velocity air at floor to condition only occupied zones; the building's single air handling unit supplies 100 percent outside air

» Operable windows provide natural ventilation to classrooms, offices and tiered lecture spaces – integrated with HVAC systems to control interior conditions

» Passive chilled beams supply cooling in conjunction with displacement ventilation systems

- » More than 50 percent of spaces receive daylighting, complemented by a series of automatic lighting controls, external shades on the east and south sides, and high efficiency luminaires in key locations
- » A planned 4,100 square foot (381 square meter) PV array on the roof will offset approximately 12 percent of the building's energy needs
- » Smart meters feed data into the building management system to record chilled water usage and lighting, mechanical and plug loads, along with temperature readouts from the air handling unit – and interface with public building dashboard visible to students, faculty and visitors

Completion Date: 2013

Ratings: LEED Platinum (LEED-NC, v.3) anticipated

PROJECT TEAM

Client/Owner: Harvey Mudd College

Glumac Design Team: MEP Engineering

Architect: Boora Architects

Design Consultants: *Civil and Structural Engineering*: KPFF Consulting Engineers; *Lighting Design*: Biella Lighting Design; *Landscape Architecture*: 2.ink Studio; *Theatre/Audio*: The Shalleck Collaborative; *Acoustics*: Jaffe Holden; *Commissioning*: CTG Energetics

General Contractor: Matt Construction Corporation

PHOTO CREDITS